Sold on Mathematics

SOLD on MATHEMATICS

Using Real Estate Ads to Teach Key Skills

by Nancy A. Silva

Peterborough, New Hampshire

Published by Crystal Springs Books
A division of Staff Development for Educators (SDE)
10 Sharon Road, PO Box 500
Peterborough, NH 03458
1-800-321-0401
www.crystalsprings.com
www.sde.com

Published 2008
Printed in the United States of America
12 11 10 09 08 1 2 3 4 5

ISBN: 978-1-934026-14-4

Library of Congress Cataloging-in-Publication Data

Silva, Nancy A., 1961-
Sold on mathematics : using real estate ads to teach key skills / by Nancy A. Silva.
p. cm.
ISBN 978-1-934026-14-4
1. Mathematics--Study and teaching--Activity programs--United States. 2. Real estate business--United States. I. Title.

QA13.S545 2008
372.7--dc22

Editor: Diane Lyons
Art Director and Designer: Soosen Dunholter
Production Coordinator: Deborah Fredericks

To my sister, Sue Ellen Rogal, and my cousins, Deb and Diane Sunderland. These strong, talented, and intelligent women have dedicated their careers to real estate law. They've inspired me more than they will ever know.

Contents

Acknowledgements

Sold on Mathematics has been 25 years in the making, and because of that there have been countless people who have offered inspiration along the way, most without even realizing it.

Thanks to my colleague Sheila May for suggesting that I get my real estate license one sunny, spring afternoon 25 years ago, after an infamous "pink slip" appeared in my school mailbox. (Read the Introduction for the full account of the story!) It was during that short-lived career that I found the spark to ignite this project.

A special thanks to my friends and next-door neighbors, Ellen Nix and Adele Wong, both award-winning agents who have sold the hundreds of houses that I never could.

Many dear friends have recently left their teaching careers to enjoy well-deserved retirements. My close friends are scattering across the country. Ellen has moved to North Carolina, Cindy has relocated to Florida, Kathy is getting adjusted to her new life in Colorado, and Karen just put her home on the market. Together we've discussed buying, selling, building, and renovating homes for hours on end, providing me with much inspiration for this book.

Throughout my 25 years at Indian Brook School in Plymouth, Massachusetts, I have been surrounded by outstanding educators. Our retiring superintendent, Barry Haskell, has kept the Plymouth Public Schools on the cutting edge, in spite of drastic budget cuts. My administrators, Jo-Ann Sikorsky, Dan Harold, Nancy Pelletier, Judd Tagg, and Dan Sylvestre have set high standards and created a child-centered school where it is a pleasure to teach. What a privilege it has been to learn from the very best! In my roles as both a classroom teacher and a mathematics specialist, I've had the opportunity to work with hundreds of students. It's been thrilling to watch them bring my ideas for this book to life. I'm also very grateful to their parents for supporting my unique approach to teaching mathematics.

Behind every great book is an outstanding editor, and I was lucky to have the very best. Diane Lyons devoted countless hours to the project, sharing her love of teaching and pushing me with her good humor. We made a great team and are thrilled with the finished product. Thanks also to Soosen Dunholter for her wonderful design work. An extra-special thanks to Sharon Smith and all of the caring, supportive folks at Crystal Springs for making this book a reality.

Finally, I must acknowledge my dad. He was always my number-one fan. Although he is not here to see *Sold on Mathematics* in print, his inspiration has touched every page.

Introduction

Welcome to *Sold on Mathematics*! I'd love to tell you that I came up with the idea for using real estate booklets to teach mathematics one lazy summer day while on vacation flipping through a real estate booklet and dreaming about purchasing a home in the Islands, but the truth is a lot less glamorous!

It was my first year of teaching, and like most of the other new teachers in my school, I got pink-slipped in the spring. I was told this procedure was strictly *pro forma*. Nervous that I might not get rehired in the fall, I followed the suggestion of a colleague and studied to become a real estate broker. After passing the exam with flying colors, I was immediately hired by a local agency.

As I spent the summer waiting for potential buyers to appear (none ever did), I had plenty of time to flip through real estate booklet after real estate booklet. As I read them, I discovered that real estate booklets are full of mathematics in action! In addition to listing prices in the hundred thousands and millions, real estate listings include fractions, percents, ordinal numbers, and decimals. They list distances measured in feet, yards, and miles. Geometry abounds! Pictures of rectangles, squares, right angles, cylinders, and isosceles triangles leap off the pages! Rather than thinking about how I could sell a house, I found myself imagining all of the different ways I could use real estate booklets to make mathematics come alive in my classroom. That summer, I had discovered a dynamic teaching tool. I also learned something about myself. I realized, more than anything else, that I wanted to be a teacher.

It's been 25 years since I closed the door on my short and not-so-sweet real estate career. But ever since then, I've been using real estate booklets to help teach students a wide range of mathematics skills. My students are always delighted when real estate booklets are used in the classroom. They're bold, dynamic, and visually attractive. I've found that since each listing is unique and cleverly written, the booklets capture and hold the interest of students at many different grade levels.

Well, I never did sell a house, but I have sold hundreds of students on the wonders of learning using real estate booklets. This book is a collection of the very best activities I've created for my students over the years. I hope you and your students enjoy them. I am thrilled to present *Sold on Mathematics*.

How to Use This Book

The only "special" material your students will need for the activities in *Sold on Mathematics* is a real estate booklet. Gathering a collection of real estate booklets is easy! Real estate booklets are free, and they're readily available in supermarkets and department stores. You'll also find that families are eager to share real estate booklets with your class. To make requesting the booklets easy, a reproducible letter for you to send home with your students is included on page 12.

It's not necessary for all of your students to use the same copy of a real estate booklet for these activities. In fact, the more resources they have to choose from the better! So go ahead and gather them from a variety of sources. When I travel, I try to pick up a real estate booklet or two to bring back to my classroom. My students have discovered that house prices and styles vary greatly from one state to the next. These out-of-town booklets are always everyone's favorites. You may wish to contact the company that distributes the real estate booklets to businesses in your area. (The address and telephone number of the company is usually listed inside the front cover of the booklet.) Rather than discarding outdated booklets, they may be willing to deliver them to your classroom. This should guarantee that you and your students will always have a fresh supply on hand.

When students first start using the real estate booklets, they may think that the listings are written in a foreign language! That's because most real estate listings include a collection of abbreviations in order to squeeze a lot of information into a very small space. Assure students that this "code" is easy and fun to crack. You may want to provide them with copies of the chart, Reading a Real Estate Listing, on page 13 to keep with their booklets.

You'll find that the activities in this book are divided into four sections.

- **5-Minute Workouts** reinforce previously taught math skills. Fun, fast, and effective, these workouts are perfect for use with the whole class. Present students with just one exercise a day and you'll keep your class mathematically fit!

5-Minute Workouts

Record Dates Used and Note Changes Made to Activities with Bold Print

Example

ADDITION ARCHERY — STRAND: NUMBER & OPERATIONS; SKILL: ADDING 1-DIGIT NUMBERS

Write the price **$395,500** on the chalkboard. Then write that price as the number sentence: **3 + 9 + 5 + 5 + 0 + 0 = 22**. Point out that the price **$395,500** has a digit sum of **22**. Challenge the class to search their real estate booklets for a house price with a digit sum of **less than 22**. List the prices and related number sentences on the chalkboard, as they are located.

✓ 9/16 ✓ 11/5 $426,900 ✓ 6/1 $537,450

ADDITION ABS — STRAND: NUMBER & OPERATIONS; SKILL: ADDING 6-DIGIT AND 7-DIGIT NUMBERS

Invite the class to read the prices of six homes to you. Use these prices to create three addition problems in working form on the chalkboard. Challenge the class to predict whether the sums will be odd or even. Invite them to predict which problem will have the greatest sum. Ask three volunteers to go to the chalkboard and solve the problems.

ADDITION ARCHERY — STRAND: NUMBER & OPERATIONS; SKILL: ADDING 1-DIGIT NUMBERS

Write the price **$395,500** on the chalkboard. Then write that price as the number sentence: **3 + 9 + 5 + 5 + 0 + 0 = 22**. Point out that the price **$395,500** has a digit sum of **22**. Challenge the class to search their real estate booklets for a house price with a digit sum of **less than 22**. List the prices and related number sentences on the chalkboard, as they are located.

ANGLE ASSISTS — STRAND: GEOMETRY; SKILL: IDENTIFYING ANGLES

Select a large photo of a featured house in a real estate booklet and ask each student to study the photo. Write the following sentence on the chalkboard: "I see a(n) ________ angle. It is ____________." Challenge the students to hunt for examples of right angles, acute angles, and obtuse angles. Invite students to share their discoveries by completing the sentence on the chalkboard. For example, "I see an acute angle. It is the peak of the roof." The students should identify the type of angle as well as where the angle is formed.

- **Scavenger Hunts** strengthen a wide assortment of mathematics skills and concepts. Students can work independently or in teams. You'll find that students are always eager to take on a new case!

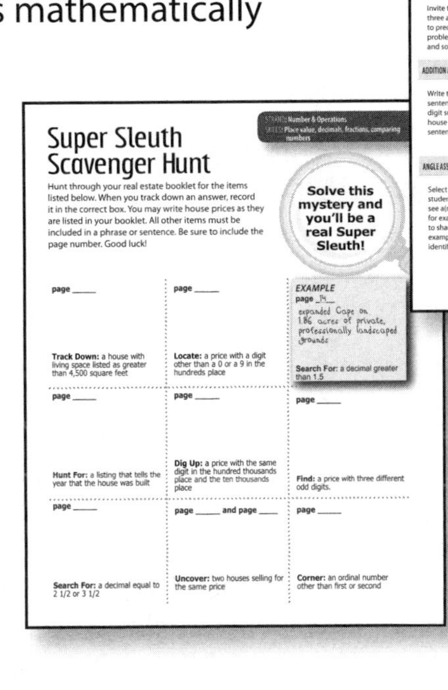
STRAND: Number & Operations
SKILLS: Place value, decimals, fractions, comparing numbers

Super Sleuth Scavenger Hunt

Hunt through your real estate booklet for the items listed below. When you track down an answer, record it in the correct box. You may write house prices as they are listed in your booklet. All other items must be included in a phrase or sentence. Be sure to include the page number. Good luck!

Solve this mystery and you'll be a real Super Sleuth!

page _____ **Track Down:** a house with living space listed as greater than 4,500 square feet	page _____ **Locate:** a price with a digit other than a 0 or a 9 in the hundreds place	*EXAMPLE* page _14_ expanded Cape on 1.86 acres of private, professionally landscaped grounds **Search For:** a decimal greater than 1.5
page _____ **Hunt For:** a listing that tells the year that the house was built	page _____ **Dig Up:** a price with the same digit in the hundred thousands place and the ten thousands place	page _____ **Find:** a price with three different odd digits.
page _____ **Search For:** a decimal equal to 2 1/2 or 3 1/2	page _____ and page _____ **Uncover:** two houses selling for the same price	page _____ **Corner:** an ordinal number other than first or second

STRAND: Number & Operations
SKILL: Ordering numbers from greatest to least

The Line Forms Here

Choose a page from your real estate booklet. Record the page number in the first column. Read the prices of the houses listed on the page. Order the house prices from greatest to least in the second column. Repeat this exercise three more times using different pages in your booklet. Then answer the questions at the bottom of the paper.

Materials:
Real estate booklet, pencil

Page Number	House prices ordered from **Greatest** to **Least**
EXAMPLE	$875,500, $875,000, $820,999, $805,799, $805,999

What is the **least** expensive house listed on this paper?

What is the **most** expensive house listed on this paper?

STRAND: Number & Operations
SKILLS: Computation, using ordinal numbers, writing numbers through the millions

The Houses That Jack Bought

Jack isn't building houses any more—he's *buying* them! Your class will enjoy putting their addition skills to the test as they each create a book inspired by the popular children's rhyme, *The House That Jack Built*. In addition to assessing the computational skills of your students, you'll also have an opportunity to evaluate their knowledge of ordinal numbers and their ability to write numbers through the millions.

Directions:

1. To kick off this activity, read some of the many book versions of *The House That Jack Built*. You may want to begin with some traditional versions and end with my favorite, the absolutely hilarious *This Is the House That Jack Built*, by Simms Taback. This entertaining, up-to-date tale has many surprises that will have your students laughing and eager to write their own books.

2. Invite students to look through their real estate booklets for houses that Jack would like to buy. Ask each student to cut out one listing and its accompanying photograph. Each student glues his cut-out pieces onto construction paper as shown in the sample below. Under the photograph, the student writes:

This is the 1st house that Jack bought. It cost ____________.
But Jack wasn't done shopping.

This is the 1st house that Jack bought. It cost $325,000. But Jack wasn't done shopping.

Each student includes the price of his house as shown. This is the first page of the student's book.

- **Skill Builders** can be used to reinforce the daily lesson or may be used to spiral back to a skill that was introduced earlier. These sheets are open-ended, which makes it easy for you to adapt them to your learners' needs.
- **Build a Book** projects are best used after your class has had much practice with the skills involved in the particular activity. These projects make great assessments or independent work.

To help you find the activities that reinforce the skills you want your students to practice, you might find the Skills Chart on pages 14–17 helpful. Activities are referenced by both the mathematics strand and the skills involved.

Parent Letter

Dear Families,

Throughout the school year, we'll be using real estate booklets in our classroom to practice and reinforce a wide variety of mathematics skills and concepts. Real estate booklets are an excellent learning resource because they feature numbers used in a wide variety of ways. They're also visually attractive and cleverly written, so they capture the interest of students. Best of all, they're free!

I'm asking for your help in gathering real estate booklets. Real estate booklets are often included as newspaper inserts. Before recycling your newspapers, please set aside any real estate inserts. You'll also find free real estate booklets in the entrance to supermarkets and department stores. I ask that you pick up a few of these booklets as you exit the store. After enjoying the booklets and viewing the homes for sale in your area, please send the booklets to school in your child's backpack. Whether traveling within our state or throughout the country, I invite you to take home a few real estate booklets for our class.

All real estate booklets are appreciated, but the class will especially welcome those that include color photographs. After looking over the real estate booklets at home, please send them into our classroom with your child. We will store them for use in future lessons.

Thank you for your help.

Sincerely,

Reading a Real Estate Listing

Real estate agents use many abbreviations in order to squeeze a lot of information into a very small space, so listings may seem like they're written in a foreign language. Relax! The code is easy and fun to crack. With the help of the chart below, you'll soon be reading real estate listings with ease.

A/C	air conditioning
AC	acres
BA	bath
BBQ	barbeque
B/I	built-in
B/I DW	built-in dishwasher
BR	bedroom
BSMT	basement
C-D-S	cul-de-sac
C-FAN	ceiling fan
CA, CAC	central air conditioning
CATH CEIL	cathedral ceiling
C/VAC	central vacuum
D/D	dishwasher/garbage disposal
DBL	double
DET	detached (garage)
DR	dining room
EIK	eat-in kitchen
F/FIN BSMT	fully finished basement
FB	full bath
FLRS	floors
FNCD YD	fenced yard
FP	fireplace
FR, F/R	family room
GAR, GRG	garage
GDOR	garage door remote
GLFP	gas log fireplace
GRMET KIT	gourmet kitchen
HB	half bath: toilet and sink
GRT RM	great room
H/W, HDWD	hardwood
HWH	hot water heater
I/G	in-ground (swimming pool)
KIT	kitchen
LA	living area
LRG	large
LR	living room
MBR	master bedroom
MICRO	microwave oven
NBD	neighborhood
NR	near
OFC	office
OH	open house
OSP	off-street parking
PL, SP, S/P	swimming pool
PVT	private
RM	room
RF	roof
SD/W	storm doors and windows
SEC	security system
SF	square feet
SPAC	spacious
STV	stove
W/D, WD	washer/dryer
W/I	walk-in (walk-in closet)
W/WW CPT	wall-to-wall carpet
WBFP	wood-burning fireplace
WD FLR	wood floor

Real estate agents continue to invent new abbreviations. As you read through your real estate booklet, you're sure to discover abbreviations not included on this chart. Be sure to list them on the back of this paper.

Sold on Mathematics Skills Chart

Activity Name	Page	Comparing Numbers	Place Value	Addition and Subtraction	Multiplication and Division	Fractions, Decimals, and Percents	Geometry	Measurement	Charts and Graphs
5-MINUTE WORKOUTS:									
Addition Abs	19	X		X					
Addition Archery	19			X					
Angle Assists	19						X		
Comparison Crunches	20	X	X						
Comparison Calf Extensions	20	X	X						
Comparison Chin-Ups	20	X			X			X	
Congruent Cartwheels	20						X		
Conversion Catches	21				X			X	
Curl-Up Comparisons	21	X	X						
Decimal Dips	21				X	X			
Decimal Dives	21	X				X			
Decimal Dribbling	21	X				X			
Digit Dropkicks	22	X		X					
Even Number Extensions	22	X							
Fraction Flips	22	X				X			
Fraction Forward Rolls	22	X				X			
Kickboxing Conversions	22				X	X			
Mental Math Martial Arts	22		X	X					
Mile March	23	X			X			X	
Odd Number Obstacle Course	23	X	X						
Ordering Olympics	23	X	X						
Ordering Overhand Throw	23	X	X						
Ordinal Offside Pass	23	X							
Perimeter Punts	24			X			X		

Activity Name	Page	Comparing Numbers	Place Value	Addition and Subtraction	Multiplication and Division	Fractions, Decimals, and Percents	Geometry	Measurement	Charts and Graphs
Place Value Pull-Ups	24	X	X						
Place Value Push-Ups	24	X	X	X					
Plane Shape Pivot	24						X		
Probability Pull-Downs	24	X				X			
Probability Pitches	25	X				X			
Rectangular Run	25				X		X		
Rounding Repetitions	25	X				X			
Rounding Rolls	25	X				X			
Rounding Rotations	25	X	X						
Rounding Rugby	26		X						
Solid Figure Strokes	26						X		
Sorting Somersaults	26	X	X						X
Square Feet Squats	26	X			X			X	
Square Inch Intervals	26	X			X			X	
Subtraction Sit-Ups	27	X	X	X					
Subtraction Splits	27	X	X	X					
Subtraction Sprints	27	X	X	X					
Tally Chart Tosses	27	X		X		X			X
Tennis Tallies	28	X		X					X
Triceps Tally	28	X	X	X					X
Waterside Workout	28				X			X	
Weighted Symmetry Stretches	28						X		
Word Name Weight Training	29	X	X						
Word Name World Series	29	X	X						
Writing Wrist Shots	29	X	X						
Zero Out Zumba	29			X					

Activity Name	Page	Comparing Numbers	Place Value	Addition and Subtraction	Multiplication and Division	Fractions, Decimals, and Percents	Geometry	Measurement	Charts and Graphs
SCAVENGER HUNTS:									
Amazing Alibi	37	X	X	X	X	X		X	
Crazy Clues	35	X	X	X	X			X	
Determined Detective	32	X	X	X	X	X			
Exciting Evidence	36	X	X	X	X	X		X	
Have a Hunch	33	X	X	X		X		X	
Marvelous Mystery	39	X		X	X				
Perfect Plot	38	X	X	X	X	X		X	
Sneaky Suspect	34	X	X	X	X			X	
Super Sleuth	31	X	X			X		X	
SKILL BUILDERS:									
Acreage Bar Graph	76	X		X	X	X		X	X
Acreage Conversions	45				X			X	
Acreage Pictograph	75	X		X	X	X		X	X
Acreage Tally Chart	74	X		X	X	X		X	X
Area Accuracy	63				X			X	
Billion-Dollar Budget	49		X	X					
Counting Bedrooms Bar Graph	70	X		X	X	X			X
Counting Bedrooms Pictograph	69	X		X	X	X			X
Counting Bedrooms Tally Chart	68	X		X	X	X			X
Decimal Delight	58					X			
Here's My Down Payment	55	X			X	X			
House Prices Bar Graph	67	X		X	X	X			X
House Prices Pictograph	66	X		X	X	X			X
House Prices Tally Chart	65	X		X	X	X			X

Activity Name	Page	Comparing Numbers	Place Value	Addition and Subtraction	Multiplication and Division	Fractions, Decimals, and Percents	Geometry	Measurement	Charts and Graphs
How Shall I Pay for This House?	53		X		X				
Just Reduced!	43		X	X					
Let's Line Up!	59	X	X		X	X			
Let's Have "Sum" Fun	48	X		X	X				
Living Space Bar Graph	73	X		X	X	X			X
Living Space Pictograph	72	X		X	X	X			
Living Space Tally Chart	71	X		X	X	X			X
Perimeter Priority	62			X				X	
Putting House Prices to the Test	56	X	X	X					
Similar or Congruent?	61						X		
Spread Out!	51	X	X						
Stop and Compare	54	X	X	X					
"Sum" Great Houses for Sale	47	X		X					
Symmetrical Search	60						X		
The Big Switch	44	X	X	X					
The Line Forms Here	42	X	X						
The Million-Dollar Switch	46	X	X	X					
The Price Is Right!	57	X	X	X	X				
The Right Angle	64						X		
This House Is a Perfect Fit!	52	X	X						
BUILD A BOOK:									
Counting on Real Estate	82	X	X			X			
Order in the House	85	X	X						
It All Adds Up	88	X	X	X					
What's the Difference?	91	X	X	X					
The Houses That Jack Bought	79	X	X	X					
Unlikely Listings	94	X	X	X	X	X		X	

How to Use the 5-MINUTE WORKOUTS

Fun and fast, these skill-maintenance exercises will keep your students mathematically fit!

Whenever your schedule has five minutes to spare, exercise your students' brains with a 5-Minute Workout. Kick off the day with Place Value Push-Ups, or start your mathematics class with Square Feet Squats. Try Decimal Dips while waiting for lunch to begin, or squeeze in Congruent Cartwheels before the final bell of the day rings. Reinforcing previously taught skills on a consistent basis is the key to student success in mathematics.

5-Minute Workouts are perfect whole-class exercises. As students work together to complete an activity, they are given the opportunity to learn from each other. These exercises also work well with small groups and can easily be adapted for one-on-one instruction.

MATERIALS: A real estate booklet is the main piece of equipment students will need for the 5-Minute Workouts. Provide your class with booklets to keep in their desks. This makes them readily available for daily use. To reduce tearing and extend the life of the booklets, have students store them in resealable plastic bags. You may wish to replace the real estate booklets every two months to keep student interest high. A chalkboard, whiteboard, or piece of chart paper is the only additional equipment that you'll need for most activities. For a few of the activities, students may also need a piece of paper and a pencil.

FOLLOW-UP: Each one of the 5-Minute Workouts is open-ended and may be repeated again and again. Some exercises ask you to put a specific example on the board. These exercises list the examples in boldface type. When you want to use these exercises for the second or third time, simply change the part of the text that has been written in boldface and you're ready to begin.

TEACHER TIP: Make a copy of the 5-Minute Workout pages and keep them in your plan book for quick reference. To keep track of the skills you've practiced with your class throughout the school year, record the dates that your class completed each workout and mark any changes made to the activities with bold print in the columns provided.

5-Minute Workouts

Record Dates Used and Note Changes Made to Activities with Bold Print

Example

ADDITION ARCHERY — STRAND: NUMBER & OPERATIONS; SKILL: ADDING 1-DIGIT NUMBERS			
Write the price **$395,500** on the chalkboard. Then write that price as the number sentence: **3 + 9 + 5 + 5 + 0 + 0 = 22**. Point out that the price **$395,500** has a digit sum of **22**. Challenge the class to search their real estate booklets for a house price with a digit sum of **less than 22**. List the prices and related number sentences on the chalkboard, as they are located.	✓ 9/15	✓ 12/5 $425,900	✓ 6/2 $537,450

ADDITION ABS

STRAND: **NUMBER & OPERATIONS**
SKILL: **ADDING 6-DIGIT AND 7-DIGIT NUMBERS**

Invite the class to read the prices of six homes to you. Use these prices to create three addition problems in working form on the chalkboard. Challenge the class to predict whether the sums will be odd or even. Invite them to predict which problem will have the greatest sum. Ask three volunteers to go to the chalkboard and solve the problems.

ADDITION ARCHERY

STRAND: **NUMBER & OPERATIONS**
SKILL: **ADDING 1-DIGIT NUMBERS**

Write the price **$395,500** on the chalkboard. Then write that price as the number sentence: **3 + 9 + 5 + 5 + 0 + 0 = 22**. Point out that the price **$395,500** has a digit sum of **22**. Challenge the class to search their real estate booklets for a house price with a digit sum of **less than 22**. List the prices and related number sentences on the chalkboard, as they are located.

ANGLE ASSISTS

STRAND: **GEOMETRY**
SKILL: **IDENTIFYING ANGLES**

Select a large photo of a featured house in a real estate booklet and ask each student to study the photo. Write the following sentence on the chalkboard: "I see a(n) ________ angle. It is ______________." Challenge the students to hunt for examples of right angles, acute angles, and obtuse angles. Invite students to share their discoveries by completing the sentence on the chalkboard. For example, "I see an acute angle. It is the peak of the roof." The students should identify the type of angle as well as where the angle is formed.

5-Minute Workouts

Record Dates Used and Note Changes Made to Activities with Bold Print

COMPARISON CRUNCHES

STRAND: **NUMBER & OPERATIONS**
SKILL: **COMPARING PRICES**

Invite the class to read the listings in their real estate booklets. Write "I'm buying a house that costs ____________________. Who is buying a house that costs more?" on the chalkboard. Choose one student to select a house and read the statement and question that you have written on the board, filling in the price of the chosen house. List the price on the chalkboard. Invite another student to answer this new question, filling in the price of a house that costs more. List the price on the chalkboard. Continue until one student has chosen the house with the greatest price. Record the number of houses purchased. Keep this information handy so that it may be referred to when this activity is repeated.

COMPARISON CALF EXTENSIONS

STRAND: **NUMBER & OPERATIONS**
SKILL: **COMPARING PRICES**

Invite the class to read the listings in their real estate booklets. Write "I'm buying a house that costs ____________________. Who is buying a house that costs less?" on the chalkboard. Choose one student to select a house and read the statement and question that you have written on the board, filling in the price of the chosen house. List the price on the chalkboard. Invite another student to answer this new question, filling in the price of a house that costs less. List the price on the chalkboard. Continue until one student has chosen the house with the lowest price. Record the number of houses purchased. Keep this information handy so that it may be referred to when this activity is repeated.

COMPARISON CHIN-UPS

STRAND: **MEASUREMENT**
SKILL: **COMPARING MEASUREMENTS IN SQUARE FEET**

Write **2,800 square feet** on the chalkboard. Challenge the class to search their real estate booklets for five houses that identify the amount of living space as less than **2,800 square feet** and five houses that identify the amount of living space as greater than **2,800 square feet**. List the ten measurements on the chalkboard.

CONGRUENT CARTWHEELS

STRAND: **GEOMETRY**
SKILL: **IDENTIFYING SIMILAR AND CONGRUENT FIGURES**

Invite the class to work in pairs. Ask the class to look at the photographs that accompany the listings in their real estate booklets. Invite the students to examine the size and shape of the doors, windows, and shutters on the houses. Challenge the students to use the vocabulary words *similar* and *congruent* to describe what they see to their partners.

5-Minute Workouts

Record Dates Used and Note Changes Made to Activities with Bold Print

CONVERSION CATCHES

STRAND: **MEASUREMENT**
SKILL: **CONVERTING ACRES TO SQUARE FEET**

Invite the class to search their real estate booklets for listings that include the label *acres.* Ask students to share the sentences they find with that label. Write each sentence on the chalkboard, underlining the number of acres and the word *acre.* Remind the class that one acre is equal to 43,560 square feet. Have the class work together to convert the number of acres in each sentence to square feet.

CURL-UP COMPARISONS

STRAND: **NUMBER & OPERATIONS**
SKILL: **COMPARING PRICES**

Write the price **$579,900** on the chalkboard. Challenge the class to search their real estate booklets for five houses selling for a price less than **$579,900** and five houses selling for a price greater than **$579,900**. List the ten prices on the chalkboard.

DECIMAL DIPS

STRAND: **NUMBER & OPERATIONS**
SKILL: **CONVERTING DECIMALS TO FRACTIONS AND MIXED NUMBERS**

Invite the class to search their real estate booklets for listings that include decimals. Ask students to share the sentences they find with decimals. Write each sentence on the chalkboard, underlining the decimal. After three sentences have been written on the chalkboard, challenge the students to convert each decimal to a fraction or mixed number. Invite three students to read the revised sentences to the class.

DECIMAL DIVES

STRAND: **NUMBER & OPERATIONS**
SKILL: **ORDERING DECIMALS FROM GREATEST TO LEAST**

Invite the class to search their real estate booklets for listings that include decimals. List the decimals students find on the chalkboard. Challenge students to work together to order the decimals from greatest to least.

DECIMAL DRIBBLING

STRAND: **NUMBER & OPERATIONS**
SKILL: **ORDERING DECIMALS FROM LEAST TO GREATEST**

Invite the class to search their real estate booklets for listings that include decimals. List the decimals students find on the chalkboard. Challenge students to work together to order the decimals from least to greatest.

5-Minute Workouts

Record Dates Used and Note Changes Made to Activities with Bold Print

DIGIT DROPKICKS

STRAND: **NUMBER & OPERATIONS**
SKILL: **ORDERING DIGITS FROM GREATEST TO LEAST**

Write the price **$997,500** on the chalkboard. Point out to the students that the digits **9**, **7**, **5**, and **0** appear in order from greatest to least. Ask the class to read the house prices in their real estate booklets. Invite the students to share any house prices made up of digits arranged from greatest to least. Record their findings on the chalkboard. Challenge a volunteer to circle the two house prices with the smallest difference.

EVEN NUMBER EXTENSIONS

STRAND: **NUMBER & OPERATIONS**
SKILL: **IDENTIFYING EVEN DIGITS**

Write the even digits 0, 2, 4, 6, and 8 on the chalkboard. Invite the class to hunt their real estate booklets for prices that include only even digits. List the prices on the chalkboard. Invite a volunteer to circle the greatest price.

FRACTION FLIPS

STRAND: **NUMBER & OPERATIONS**
SKILL: **ORDERING FRACTIONS FROM GREATEST TO LEAST**

Invite the class to search their real estate booklets for listings that include fractions. List the fractions on the chalkboard. Challenge the class to work together to order the fractions from greatest to least.

FRACTION FORWARD ROLLS

STRAND: **NUMBER & OPERATIONS**
SKILL: **ORDERING FRACTIONS FROM LEAST TO GREATEST**

Invite the class to search their real estate booklets for listings that include fractions. List the fractions on the chalkboard. Challenge the class to work together to order the fractions from least to greatest.

KICKBOXING CONVERSIONS

STRAND: **NUMBER & OPERATIONS**
SKILL: **CONVERTING FRACTIONS AND MIXED NUMBERS TO DECIMALS**

Invite the class to search their real estate booklets for listings that use fractions or mixed numbers. Ask students to share the sentences they find with fractions or mixed numbers. Write each sentence on the chalkboard, underlining the fractions or mixed numbers. After three sentences have been written on the chalkboard, challenge the students to convert each fraction or mixed number to a decimal. Invite three students to read the revised sentences to the class.

MENTAL MATH MARTIAL ARTS

STRAND: **NUMBER & OPERATIONS**
SKILL: **SUBTRACTION WITH TRADING**

Invite the class to read six house prices to you from their real estate booklets. List the prices on the chalkboard. Challenge the class to use mental math to find what the new house prices will be if the cost of each house is reduced by **$10,000.**

5-Minute Workouts

Record Dates Used and Note Changes Made to Activities with Bold Print

MILE MARCH

STRAND: **MEASUREMENT**
SKILL: **CONVERTING MILES TO FEET**

Invite the class to search their real estate booklets for listings that include the label *miles*. Ask students to share the sentences they find with that label. Write each sentence on the chalkboard, underlining the number of miles and the word *miles*. Remind the class that one mile is equal to 5,280 feet. Have the class work together to convert the number of miles in each sentence to feet.

ODD NUMBER OBSTACLE COURSE

STRAND: **NUMBER & OPERATIONS**
SKILL: **IDENTIFYING ODD NUMBERS**

Write the odd digits 1, 3, 5, 7, and 9 on the chalkboard. Invite the class to hunt for prices that have an odd digit in the hundred thousands place, the ten thousands place, and the thousands place. List five of the prices students find on the chalkboard. Invite a volunteer to underline the digit in the hundred thousands place in each price. Ask another student to circle the digit in the thousands place in all five prices.

ORDERING OLYMPICS

STRAND: **NUMBER & OPERATIONS**
SKILL: **ORDERING NUMBERS FROM LEAST TO GREATEST**

Invite the class to read the house prices in their real estate booklets. Ask six students to each share the price of their favorite house. Challenge volunteers to arrange the house prices from least to greatest.

ORDERING OVERHAND THROW

STRAND: **NUMBER & OPERATIONS**
SKILL: **ORDERING NUMBERS FROM GREATEST TO LEAST**

Invite the class to read the house prices in their real estate booklets. Ask six students to each share the price of their favorite house. Challenge volunteers to arrange the house prices from greatest to least.

ORDINAL OFFSIDE PASS

STRAND: **NUMBER & OPERATIONS**
SKILL: **IDENTIFYING ORDINAL NUMBERS**

Write the ordinal number **fifth** and the abbreviation **5th** on the chalkboard. Invite the class to search their real estate booklets for listings that include either an ordinal number word name or the abbreviation for an ordinal number. List the word names or abbreviations on the chalkboard. For each ordinal number word name that is found, invite a student to write the abbreviation on the chalkboard. For each ordinal number abbreviation that is located, challenge a student to write the ordinal number word name on the chalkboard.

5-Minute Workouts

Record Dates Used and Note Changes Made to Activities with Bold Print

PERIMETER PUNTS

STRAND: **MEASUREMENT**
SKILL: **FINDING THE PERIMETER OF A RECTANGLE**

Invite the class to search their real estate booklets for listings that include the dimensions for a rectangular swimming pool or deck. Invite students to draw the rectangular pools or decks they find on the chalkboard and label the dimensions. Remind students that perimeter = length + length + width + width. Challenge the class to find the perimeter of each swimming pool or deck.

PLACE VALUE PULL-UPS

STRAND: **NUMBER & OPERATIONS**
SKILL: **IDENTIFYING PLACE VALUE RELATIONS**

Invite the class to read the house prices in their real estate booklets. Ask six students to each share the price of their favorite house. List the prices on the chalkboard. Invite a volunteer to circle the digit in the hundred thousands place in each price. Ask another volunteer to underline the digit in the thousands place in each price. Challenge the class to identify the new price of each house if the digit in the hundred thousands place is switched with the digit in the thousands place.

PLACE VALUE PUSH-UPS

STRAND: **NUMBER & OPERATIONS**
SKILL: **IDENTIFYING PLACE VALUE RELATIONS**

Write the price **$1,949,900** on the chalkboard. Point out that there is a **nine** in the hundred thousands place, the thousands place, and the hundreds place. Ask the class to find the value of each **nine** (900,000, 9,000, and 900) and the combined value of the **nines** (909,900). Challenge students to locate prices in their real estate booklets with two or more **fives** in the price. Record their findings on the chalkboard. Ask the class to find the value of each **five** and the combined value of the **fives**.

PLANE SHAPE PIVOT

STRAND: **GEOMETRY**
SKILL: **IDENTIFYING PLANE SHAPES**

Review the names and properties of plane shapes with your students. Write the sentence, "I see an example of a _____. It is a ______." Invite the class to search the photos in their real estate booklets for examples of circles, squares, rectangles, hexagons, and other plane shapes. Ask the students to share their discoveries by completing the sentence on the chalkboard. For example, "I see an example of a rectangle. It is a shutter." Invite the rest of the class to look through their booklets to see if they can spot additional examples of the given plane shape.

PROBABILITY PULL-DOWNS

STRAND: **NUMBER & OPERATIONS, DATA ANALYSIS & PROBABILITY**
SKILLS: **EXPRESSING PROBABILITY IN THE FORM OF A FRACTION**

Invite the class to turn to any page in their real estate booklets. Allow a few minutes for the students to read through the listings to become familiar with their features. Ask the students to express, as a fraction, the probability of finding a listing for a house that **has more than three bedrooms.** Invite the class to share their fractions. Challenge the class to sort the fractions into three groups: less than 1/4, equal to 1/4, or greater than 1/4.

5-Minute Workouts

Record Dates Used and Note Changes Made to Activities with Bold Print

PROBABILITY PITCHES

STRAND: **NUMBER & OPERATIONS, DATA ANALYSIS & PROBABILITY**
SKILLS: **EXPRESSING PROBABILITY IN THE FORM OF A FRACTION**

Invite the class to turn to any page in their real estate booklets. Allow a few minutes for the students to examine the photographs that accompany the listings to become familiar with their features. Ask the students to express as a fraction the probability of finding a photograph that **has an even number of windows on the front.** Invite the students to share their fractions. Challenge the class to sort the fractions into three groups: less than 1/2, equal to 1/2, or greater than 1/2.

RECTANGULAR RUN

STRAND: **GEOMETRY**
SKILL: **IDENTIFYING THE DIMENSIONS OF A RECTANGLE GIVEN THE AREA**

Invite the class to hunt for a real estate listing that identifies the square footage of the home. Write the area on the chalkboard. Have the class work together to design two or more different squares or rectangles with that area. Label the dimensions. (For example, if the area of a home is 2,400 square feet, the student might design a home that is 80 x 30 feet or 60 x 40 feet.)

ROUNDING REPETITIONS

STRAND: **NUMBER & OPERATIONS**
SKILL: **ROUNDING DECIMALS TO THE NEAREST TENTH**

Write **1.45** acres on the chalkboard. Review rounding rules. Explain and show how **1.45** rounds to **1.5** when rounded to the nearest tenth. Invite the class to search the listings in their real estate booklets for decimals. The decimals should include both the tenths place and the hundredths place. List the decimals on the chalkboard. Challenge the students to work together to round each decimal to the nearest tenth. Ask a volunteer to circle the decimals greater than **1.5**.

ROUNDING ROLLS

STRAND: **NUMBER & OPERATIONS**
SKILL: **ROUNDING DECIMALS TO THE NEAREST WHOLE NUMBER**

Write **2.7** acres on the chalkboard. Review rounding rules. Explain and show how **2.7** rounds to **3** when rounded to the nearest whole number. Invite the class to search the listings in their real estate booklets for decimals. The decimals need only include digits in the tenths place. List the decimals on the chalkboard. Challenge the students to work together to round each decimal to the nearest whole number. Ask a volunteer to circle the decimals that round to a whole number greater than **3**.

ROUNDING ROTATIONS

STRAND: **NUMBER & OPERATIONS**
SKILL: **ROUNDING PRICES TO THE NEAREST HUNDRED THOUSAND**

Write the prices **$559,900** and **$625,900** on the chalkboard. Review rounding rules with the class. Explain that **$559,900** and **$625,900** both round to **$600,000** when rounded to the nearest hundred thousand. Invite students to hunt for prices in their real estate booklets that round to **$600,000** when rounded to the nearest hundred thousand.

5-Minute Workouts

Record Dates Used and Note Changes Made to Activities with Bold Print

ROUNDING RUGBY

STRAND: **NUMBER & OPERATIONS**
SKILL: **ROUNDING PRICES TO THE NEAREST TEN THOUSAND**

Write the prices **$285,500** and **$290,000** on the chalkboard. Review rounding rules with the class. Explain that **$285,500** rounds to **$290,000** when rounded to the nearest ten thousand. Invite students to share five prices found in their real estate booklets. Have the class work together to round the prices to the nearest ten thousand.

SOLID FIGURE STROKES

STRAND: **GEOMETRY**
SKILL: **IDENTIFYING SOLID FIGURES**

Review the names and properties of solid figures with your students. Write the sentence, "I see an example of a _______. It is a _____________." Invite the class to search the photos in their real estate booklets for examples of cones, cylinders, rectangular prisms, and other solid figures. Ask the students to share their discoveries by completing the sentence on the chalkboard. For example, "I see a cylinder. It is the post below the mailbox." Invite the rest of the class to look through their booklets to see if they can spot additional examples of the given solid figure.

SORTING SOMERSAULTS

STRAND: **NUMBER & OPERATIONS, DATA ANALYSIS & PROBABILITY**
SKILL: **CREATING A TALLY CHART**

Write the following three categories on the chalkboard: **Less Than $500,000**, **Between $500,000 and $1,000,000**, and **Greater Than $1,000,000**. Divide the class into pairs or small groups. Invite students to select two or more pages in their real estate booklets. Ask each group to record the number of house prices for each of the three categories on a piece of paper. Invite a representative from each group to transfer this information to the chalkboard using tally marks. Have the class work together to total the number of tally marks in each category. Discuss the results.

SQUARE FEET SQUATS

STRAND: **MEASUREMENT**
SKILL: **CONVERTING SQUARE FEET TO SQUARE YARDS**

Invite the class to search their real estate booklets for listings that include the label *square feet*. Ask the students to share the sentences that include the label. Write the sentences on the chalkboard, underlining the number of square feet and the words *square feet*. Remind students that 9 square feet is equal to 1 square yard. Have the class work together to convert the area from square feet to square yards.

SQUARE INCH INTERVALS

STRAND: **MEASUREMENT**
SKILL: **CONVERTING SQUARE FEET TO SQUARE INCHES**

Invite the class to search their real estate booklets for listings that include the label *square feet*. Ask the students to share the sentences that include the label. Write the sentences on the chalkboard, underlining the number of square feet and the words *square feet*. Remind students that 1 square foot is equal to 144 square inches. Have the class work together to convert the area from square feet to square inches.

5-Minute Workouts

Record Dates Used and Note Changes Made to Activities with Bold Print

SUBTRACTION SIT-UPS

STRAND: **NUMBER & OPERATIONS**
SKILLS: **SUBTRACTION WITH TRADING, USING ADDITION TO CHECK SUBTRACTION**

Invite the class to read six house prices to you from their real estate booklets. List the prices on the chalkboard. Explain to the class that the selling prices have just been reduced by **$9,550**. Invite volunteers to come to the chalkboard to complete the subtraction needed to find the new selling price for each house. Ask students at their seats to use addition to check the accuracy of their classmates' subtraction.

SUBTRACTION SPLITS

STRAND: **NUMBER & OPERATIONS**
SKILLS: **SUBTRACTION WITH TRADING, ESTIMATING DIFFERENCES**

Invite the class to read six house prices to you from their real estate booklets. Use these prices to create three subtraction problems in working form on the chalkboard. Challenge the class to predict whether the differences will be odd or even and ask them to predict which problem will have the greatest difference. Invite three volunteers to go to the chalkboard and solve the problems.

SUBTRACTION SPRINTS

STRAND: **NUMBER & OPERATIONS**
SKILL: **SUBTRACTING 6-DIGIT AND 7-DIGIT NUMBERS**

Invite the class to search their real estate booklets for pairs of house prices that have a price difference of as close to **$50,000** as possible. As the students share the price pairs, create subtraction problems in working form on the chalkboard. Choose volunteers to solve the problems. Circle the problem with the price difference closest to **$50,000**.

TALLY CHART TOSSES

STRAND: **DATA ANALYSIS & PROBABILITY**
SKILL: **CREATING A TALLY CHART**

List the following four categories on the chalkboard: **Less Than 2 bathrooms**, **2 or 2.5 bathrooms**, **3 or 3.5 bathrooms**, and **More Than 3.5 bathrooms**. Divide the class into pairs or small groups. Assign each pair or group two or more pages in the real estate booklet. The number of pages assigned will vary depending on the size of the real estate booklet and the number of students in the class. Ask each group to record the number of bathrooms identified in each listing found on their pages. Invite a representative from each group to transfer this information to the chalkboard using tally marks. Have the class work together to identify the total number of tally marks in each category. Discuss the results.

5-Minute Workouts

Record Dates Used and Note Changes Made to Activities with Bold Print

TENNIS TALLIES

STRAND: **DATA ANALYSIS & PROBABILITY**
SKILL: **CREATING A PICTOGRAPH**

List the following six categories on the chalkboard: **1 Bedroom**, **2 Bedrooms**, **3 Bedrooms**, **4 Bedrooms**, and **5 or More Bedrooms**. Divide the class into pairs or small groups. Assign each pair or group two or more pages in the real estate booklet. The number of pages assigned will vary depending on the size of the real estate booklet and the number of students in the class. Ask each group to record the number of bedrooms identified in each listing found on their pages. Invite a representative from each group to transfer this information to the chalkboard using circles. Explain to the students that one circle will represent two real estate listings. Have the class work together to total the number of symbols in each category. Discuss the results. Challenge the students to identify five questions that may be answered by reading the pictograph.

TRICEPS TALLY

STRAND: **DATA ANALYSIS & PROBABILITY**
SKILL: **CREATING A PICTOGRAPH**

List the following three categories on the chalkboard: **Less Than $500,000**, **Between $500,000 and $1,000,000**, and **Greater Than $1,000,000**. Divide the class into pairs or small groups. Assign each pair or group two or more pages in the real estate booklet. The number of pages assigned will vary depending on the size of the real estate booklet and the number of students in the class. Ask each group to record the number of house prices for each of the three categories on a piece of paper. Invite a representative from each group to transfer this information to the chalkboard using squares. Explain to the students that one square will represent four real estate listings. Have the class work together to identify the total number of tally marks in each category. Discuss the results. Challenge the students to make five true statements about the tally chart.

WATERSIDE WORKOUT

STRAND: **NUMBER & OPERATIONS**
SKILL: **WRITING WORD NAMES FOR NUMBERS**

Invite the class to search their real estate booklets for listings that include the dimensions of a rectangular swimming pool or deck. Invite students to draw the rectangular pools or decks they find on the chalkboard and label the dimensions. Remind students that area = length x width. Challenge the class to find the area of each swimming pool or deck.

WEIGHTED SYMMETRY STRETCHES

STRAND: **GEOMOTRY**
SKILL: **LOCATING SYMMETRICAL ITEMS**

Ask the class to look at the photographs that accompany each listing in their real estate booklets. Challenge the class to locate examples of symmetry. The students may choose to focus on a whole house, part of a house, or items found in the yard. Invite students to share their discoveries by drawing the items on the chalkboard. Invite volunteers to add dashed lines of symmetry to the drawings to prove that they are symmetrical.

5-Minute Workouts

Record Dates Used and Note Changes Made to Activities with Bold Print

WORD NAME WEIGHT TRAINING

STRAND: **MEASUREMENT**
SKILL: **FINDING THE AREA OF A RECTANGLE**

Ask students to read through the listings in their real estate booklets and choose their favorite house. Invite five students to come to the chalkboard and write the word name for the price of their favorite house. Read the word names for each price together. Call on volunteers to write the prices in standard form beside each word name. Challenge the rest of the class to search through their real estate booklets for additional houses selling for the five prices listed on the chalkboard.

WORD NAME WORLD SERIES

STRAND: **NUMBER & OPERATIONS**
SKILL: **MATCHING WORD NAMES WITH NUMBERS**

Write the following five word name parts on the chalkboard: **1. two million**, **2. five hundred**, **3. fifty thousand**, **4. four hundred thousand**, and **5. ninety-five thousand**. Invite the class to search their real estate booklets for one or more house prices that match each word name part. List the prices on the chalkboard. Invite a volunteer to write the word name beside each price.

WRITING WRIST SHOTS

STRAND: **NUMBER & OPERATIONS**
SKILL: **MATCHING WORD NAMES WITH NUMBERS WRITTEN IN STANDARD FORM**

Write the price **$775,000** on the chalkboard. Then write the price in expanded form: **700,000 + 70,000 + 5,000**. Ask students to share house prices from their real estate booklets. List five prices on the chalkboard. Invite volunteers to come to the chalkboard and write the house prices in expanded form.

ZERO OUT ZUMBA

STRAND: **NUMBER & OPERATIONS**
SKILL: **WRITING NUMBERS IN EXPANDED FORM**

Explain to the class that they have been given a billion-dollar budget and may purchase as many homes as possible without going over budget. Allow students time to read the listings in their real estate booklets. To begin, invite a student to share the price of his favorite home. Have students work together to subtract the prices of the home from one billion dollars. Continue by having students select more homes, subtracting the house prices until there is not enough money in the billion-dollar budget to make any additional purchases. Record the number of houses purchased and the amount of money left. Keep this information handy so that it may be referred to when this activity is repeated.

How to Use the SCAVENGER HUNTS

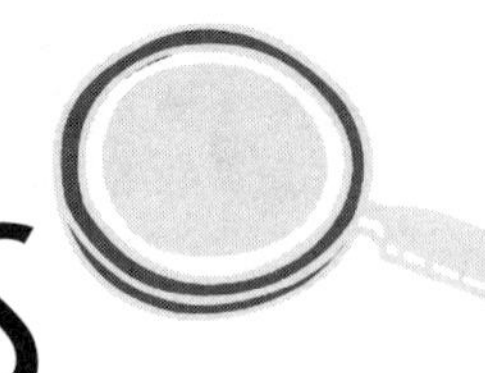

These hunts will have your super sleuths racing through their real estate booklets, eagerly searching for examples of mathematics in action!

There are so many different ways to use the Scavenger Hunts with your students. You may choose to have your class work on a hunt independently, in pairs, or in small teams. Your class may work until one student or group has located all of the requested items, or you may announce, "Case closed," after 20 to 40 minutes have passed. Your students will love the game element of this activity! The Scavenger Hunts are sure to keep your students engaged, so you may want to assign a hunt to half of your class while you provide small group instruction to the other half.

MATERIALS: The real estate booklet, a copy of the scavenger hunt, and a pencil are the only tools your detectives will need. Students may use any real estate booklets that are available; however, if your class is playing until one detective finds all of the requested items, I have found that it's best—or as my students tell me, fair—if everyone is using the same resource. Each Scavenger Hunt sheet can be reused throughout the school year as new batches of real estate booklets are introduced.

FOLLOW-UP: When a scavenger hunt has come to a close, invite the detectives to share the "evidence" they located. For example, if the students were asked to find an ordinal number other than first or second, list *all* of the ordinal numbers that were discovered on the chalkboard or whiteboard. Ask a volunteer to circle the greatest ordinal number. Challenge another student to write the abbreviation for one of the ordinal number words. This may be the perfect time for you to conduct a mini-lesson or do a quick review of a previously taught skill. You may choose to have each student verify a fellow classmate's scavenger hunt or collect the papers and look them over yourself.

TEACHER TIP: A blank scavenger hunt sheet has been provided for you on page 40. Use it to create hunts that address a particular skill your class is working on, or to make personalized hunts for learners who need modifications or challenges. You'll find that after your students become familiar with the many ways that numbers are used in real estate booklets, they'll be eager to create their own hunts for classmates to solve!

STRAND: **Number & Operations**
SKILLS: **Place value, decimals, fractions, comparing numbers**

Super Sleuth Scavenger Hunt

Hunt through your real estate booklet for the items listed below. When you track down an answer, record it in the correct box. You may write house prices as they are listed in your booklet. All other items must be included in a phrase or sentence. Be sure to include the page number. Good luck!

Solve this mystery and you'll be a real Super Sleuth!

page ______ **Track Down:** a house with living space listed as greater than 4,500 square feet	page ______ **Locate:** a price with a digit other than a 0 or a 9 in the hundreds place	***EXAMPLE*** **page** _14_ expanded Cape on 1.86 acres of private, professionally landscaped grounds **Search For:** a decimal greater than 1.5
page ______ **Hunt For:** a listing that tells the year that the house was built	page ______ **Dig Up:** a price with the same digit in the hundred thousands place and the ten thousands place	page ______ **Find:** a price with three different odd digits.
page ______ **Search For:** a decimal equal to 2 1/2 or 3 1/2	page ______and page ____ **Uncover:** two houses selling for the same price	page ______ **Corner:** an ordinal number other than first or second

STRAND: **Number & Operations**
SKILLS: **Place value, decimals, fractions, comparing numbers**

Determined Detective Scavenger Hunt

Hunt through your real estate booklet for the items listed below. When you track down an answer, record it in the correct box. You may write house prices as they are listed in your booklet. All other items must be included in a phrase or sentence. Be sure to include the page number. Good luck!

Stay determined—you can crack this case!

page ______ and page ____ **Find:** two different prices that round to $600,000 when rounded to the nearest hundred thousand	page ______ **Uncover:** a price with the same digit in the ten thousands place and the hundreds place	*EXAMPLE* page 38 $499,000 **Uncover:** a price that rounds to $500,000 when rounded to the nearest hundred thousand
page ______ **Dig Up:** a price between $419,500 and $491,500	page ______ **Corner:** a house that has between 3,000 and 4,000 square feet of living space	page ______ **Track Down:** a house with exactly twice as many bedrooms as bathrooms
page ______ **Search For:** a decimal less than .5	page ______ **Hunt For:** a mixed number equal to 1.5	page ______ **Locate:** a decimal greater than 1.75

Have a Hunch Scavenger Hunt

STRANDS: **Number & Operations, Measurement**
SKILLS: **Place value, decimals, fractions, comparing numbers**

Hunt through your real estate booklet for the items listed below. When you track down an answer, record it in the correct box. You may write house prices as they are listed in your booklet. All other items must be included in a phrase or sentence. Be sure to include the page number. Good luck!

Trust your hunches! That's what great detectives do!

page _____ **Uncover:** a price between \$610,000 and \$675,000	page _____ and page ____ **Locate:** two different prices that round to \$500,000 when rounded to the nearest hundred thousand	*EXAMPLE* page _23__ early 20th century home replete with tin ceilings **Hunt For:** an even ordinal number greater than fourth
page _____ **Dig Up:** a listing that identifies the dimensions of the master bedroom	page _____ and page ____ **Find:** two different prices with a 7 in the ten thousands place	page _____ and page ____ **Search For:** two different prices with a difference of less than \$25,000
page _____ **Hunt for:** a mixed number greater than 3 1/2	page _____ **Corner:** an odd ordinal number greater than fifth	page _____ **Track Down:** a price with a digit sum of exactly 20

Sneaky Suspect Scavenger Hunt

STRANDS: **Number & Operations, Measurement**
SKILLS: **Place value, decimals, fractions, comparing numbers, measuring area and distance**

Hunt through your real estate booklet for the items listed below. When you track down an answer, record it in the correct box. You may write house prices as they are listed in your booklet. All other items must be included in a phrase or sentence. Be sure to include the page number. Good luck!

You can find the suspect hiding in your real estate booklet!

page ______ **Find:** a price with an even number in the hundred thousands place and the thousands place	page ______ **Track Down:** a price that has a digit sum less than 10	*EXAMPLE* page _53_ . . . 2 baths, deck, full basement, and new septic. **Uncover:** a house with an even number of bathrooms
page ______ **Search For:** a listing that identifies the dimensions of the master bedroom	page ______ and page _____ **Uncover:** two prices with a difference between $1,000 and $8,000	page ______ **Hunt For:** a listing that identifies the dimensions of an in-ground swimming pool
page ______ **Locate:** a house with an odd number of bedrooms	page ______ and page _____ **Dig Up:** two different prices made up of the same digits	page ______ **Corner:** a listing that includes the label *acres*

Crazy Clues Scavenger Hunt

STRANDS: **Number & Operations, Measurement**
SKILLS: **Place value, comparing numbers, measuring area and distance**

Hunt through your real estate booklet for the items listed below. When you track down an answer, record it in the correct box. You may write house prices as they are listed in your booklet. All other items must be included in a phrase or sentence. Be sure to include the page number. Good luck!

Don't go crazy! All the clues you need are in your real estate booklet.

page ______ **Uncover:** a price with an odd number in the hundred thousands place and the thousands place	page ______ **Locate:** a price with ascending digits in the hundred thousands place, ten thousands place, and thousands place	*EXAMPLE* page 7 $975,000 **Surround:** a price with descending digits in the hundred thousands, ten thousands, and thousands place
page ______ **Corner:** a house with an even number of bedrooms	page ______ **Hunt For:** a price that has a digit sum greater than 25	page ______, page ______, page ______, and page ______ **Search For:** four prices that together include each digit (0, 1, 2, 3, 4, 5, 6, 7, 8, 9) at least once
page ______ **Find:** a listing that identifies the dimensions of a deck	page ______ **Dig Up:** a listing that includes the label miles	page ______ **Track Down:** a listing that includes the label *square feet*

Exciting Evidence Scavenger Hunt

STRANDS: **Number & Operations, Measurement**
SKILLS: **Place value, computation, comparing numbers, measuring area, distance, and time**

Hunt through your real estate booklet for the items listed below. When you track down an answer, record it in the correct box. You may write house prices as they are listed in your booklet. All other items must be included in a phrase or sentence. Be sure to include the page number. Good luck!

Hunting down evidence is exciting !

page _____ **Corner:** a listing that identifies the age of the house but not the year it was built	page _____ **Search For:** a price that has a digit sum between 24 and 30	*EXAMPLE* page 27 \$875,000 \$949,500 **Find:** two different prices that each round to \$900,000 when rounded to the nearest hundred thousand
page _____ , page _____ , and page _____ **Track Down:** three different prices made up of the digits 0, 5, 7, or 9	page _____ **Locate:** a listing with half as many bathrooms as bedrooms	page _____ and page _____ **Surround:** two prices with a difference greater than \$1,000,000
page _____ **Hunt For:** a Roman numeral	page _____ **Uncover:** a price with three pairs of matching digits	page _____ **Dig Up:** a listing that includes the label *minutes*

Amazing Alibi Scavenger Hunt

STRANDS: **Number & Operations, Measurement**
SKILLS: **Place value, computation, comparing numbers, measuring area and distance**

Hunt through your real estate booklet for the items listed below. When you track down an answer, record it in the correct box. You may write house prices as they are listed in your booklet. All other items must be included in a phrase or sentence. Be sure to include the page number. Good luck!

You won't need an alibi. The facts are all in your real estate booklet!

page ______ **Track Down:** a listing that includes the label *feet*, but not *square feet*	page ______ **Locate:** a listing with a greater number of bathrooms than bedrooms	*EXAMPLE* page _14_ $379,900 3 + 7 + 9 + 9 + 0 + 0 = 28 **Search For:** a price with an even digit sum between 20 and 30
page ______ **Corner:** a listing that identifies the amount of frontage on a lake, a pond, a river, or the ocean	page ______ **Hunt For:** a listing that includes the dimensions of a barn or garage	page ______ and page ______ **Find:** two different decimals that round to 2 when rounded to the nearest whole number
page ______ **Surround:** a price with an odd digit sum greater than 30	page ______ **Search For:** a price that will include 5,000 when written in expanded form	page ______ **Locate:** an ordinal number other than first, second, third, or fourth

Perfect Plot Scavenger Hunt

STRANDS: **Number & Operations, Measurement, Geometry**
SKILLS: **Place value, computation, identifying odd and even numbers, measuring in feet**

Hunt through your real estate booklet for the items listed below. When you track down an answer, record it in the correct box. You may write house prices as they are listed in your booklet. All other items must be included in a phrase or sentence. Be sure to include the page number. Good luck!

Help to uncover the "perfect plot!"

page ______

Find: a house that can be purchased with between 77,777 and 99,999 ten-dollar bills

page ______

Track Down: a listing with an odd number of bedrooms and an even number of bathrooms

EXAMPLE
page 91

$609,000
6 + 9 = 15
$325,500
(3 + 2) + (5 + 5) = 15

Locate: two different prices with a digit sum of 15

page ______

Corner: a price that would include 50,000 when written in expanded form

page ______

Uncover: a price with an odd number in the ten thousands place and an even number in the hundred thousands place

page ______ , page ______ , and page ______

Locate: three different house prices with the same digit sum

page ______

Hunt For: the dimensions of a square master bedroom

page ______

Surround: a listing that identifies a ceiling height greater than 8 feet

page ______

Search For: a listing that includes the word *half* or *quarter*

STRAND: **Number & Operations**
SKILLS: **Computation, identifying even and odd numbers, comparing sums, identifying multiples of 3, 5, and 10**

Marvelous Mystery Scavenger Hunt

Calling all detectives! Can you solve this marvelous mystery?

Hunt through your real estate booklet for the items listed below. When you track down an answer, record it in the correct box. You may write house prices as they are listed in your booklet. All other items must be included in a phrase or sentence. Be sure to include the page number. Good luck!

page ______ **Dig Up:** a price with a digit sum greater than 25	page ______ **Hunt For:** a price with a digit sum that is a multiple of 3	*EXAMPLE* page _14_ \$712,000 (7 + 1) + 2 = 10 \$325,000 (3 + 2) + 5 = 10 **Locate:** two different prices with a digit sum of 10
page ______ **Search For:** a price with a digit sum less than 6	page ______ **Locate:** a price with an even digit sum greater than 14 but less than 24	page _25_ **Find:** a price with a digit sum that is a multiple of 5 but not a multiple of 10
page ______ and page ______ **Uncover:** two different prices with the same odd digit sum	page ______ , page ______ , and page ______ **Track Down:** three prices with a combined digit sum greater than 50	page ______ and page ______ **Surround:** two prices with a combined digit sum of 30

Just for You Scavenger Hunt

Hunt through your real estate booklet for the items listed below. When you track down an answer, record it in the correct box. You may write house prices as they are listed in your booklet. All other items must be included in a phrase or sentence. Be sure to include the page number. Good luck!

Here's a scavenger hunt created for you by ______________!

page _____ **Hunt For:** ________________	page _____ **Search For:** ________________	page _____ and page _____ **Uncover:** ________________
page _____ **Track Down:** ________________	page _____ **Locate:** ________________	page _____ and page _____ **Hunt For:** ________________
page _____ **Uncover:** ________________	page _____ **Surround:** ________________	page _____ **Find:** ________________

How to Use the SKILL BUILDERS

Students will love these open-ended activity sheets. They're a refreshing alternative to traditional worksheets!

When and how you use the Skill Builders will depend on your needs. You might assign one to reinforce the daily lesson. Or, you may choose to use one that spirals back to a skill that was introduced weeks or months earlier in an effort to maintain mastery. Skill Builders make excellent independent work and are perfect for small group remediation. They also serve well as assessment tools.

On average, Skill Builders take between 20 and 40 minutes to complete. However, the time will vary depending on the difficulty of the skills addressed and the ability level of your students.

MATERIALS: To complete most of the Skill Builders, all your students will need are their real estate booklets and a pencil. For the rest of the activities, you'll need to have crayons, highlighters, glue, scissors, and construction paper on hand. You'll find a materials list included on each activity sheet. Since the Skill Builders are open-ended, they can be repeated again and again throughout the school year as new batches of real estate booklets are introduced.

FOLLOW-UP: Skill Builders are open-ended, so students will select listings from their own booklets to complete each task. When an activity sheet is finished, invite students to compare their sheets and discuss their unique choices with classmates. As students talk about their work, you'll have the opportunity to hear how key vocabulary words are used, and you'll be able to observe your students' level of understanding for the given skill. This may be the perfect time for you to conduct a mini-lesson or do a quick review of a previously taught skill.

TEACHER TIP: In order for all students to experience success, you may choose to differentiate your requirements. Invite high-achieving students to complete additional examples on a separate sheet of paper. You may also ask these students to write about their completed Skill Builder on the back of the sheet using mathematics vocabulary. To modify the Skill Builders for students who need more support, cut out listings in advance and/or highlight the key numbers or words featured on a page. You may also reduce the number of examples these students are asked to complete.

The Line Forms Here

STRAND: **Number & Operations**
SKILL: **Ordering numbers from greatest to least**

Choose a page from your real estate booklet. Record the page number in the first column. Read the prices of the houses listed on the page. Order the house prices from greatest to least in the second column. Repeat this exercise three more times using different pages in your booklet. Then answer the questions at the bottom of the paper.

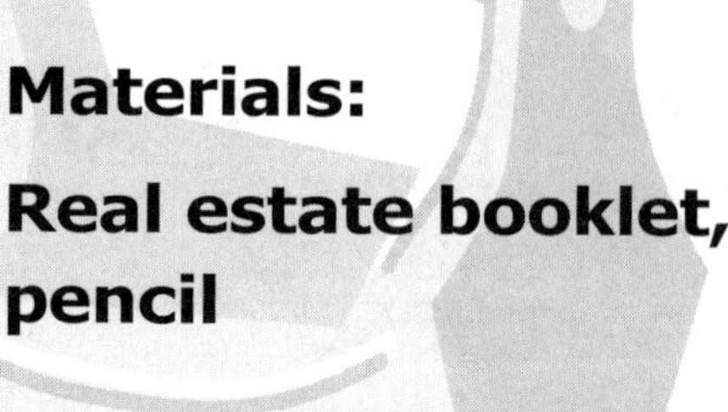

Materials:

Real estate booklet, pencil

Page Number	House prices ordered from **Greatest** to **Least**
EXAMPLE	$875,500, $875,000, $820,999, $805,799, $805,999

What is the **least** expensive house listed on this paper? ______________________

What is the **most** expensive house listed on this paper? ______________________

Just Reduced!

STRAND: **Number & Operations**
SKILL: **Subtraction with and without trading**

Select a house from your real estate booklet. Record the page number and the current selling price of the house. Subtract to find what the price of the house will be if the original price is reduced by $5,000, by $10,000, and by $25,000. Do your work on a separate piece of paper. Record the new prices in the appropriate columns. Repeat these steps for six more houses. The original selling price of each house must be different. Staple your work paper to this sheet when you have completed the assignment.

Materials:

Real estate booklet, work paper, pencil

Page Number	Current House Price	$5,000 Price Reduction	$10,000 Price Reduction	$25,000 Price Reduction
EXAMPLE	$382,500	$377,500	$372,500	$357,500

The Big Switch

STRAND: **Number & Operations**
SKILLS: **Place value, computing differences**

Choose a house from your real estate booklet. Record the page number and the current selling price. Underline the digit in the thousands place and the digit in the hundred thousands place. Exchange the underlined digits and record the new price in the third column. Find the difference between the original price and the new price. Record the difference in the last column. Indicate whether the difference is an increase or a decrease. Repeat these steps for seven more houses. Staple your work paper to this sheet when you have completed the assignment.

Materials:

Real estate booklet, work paper, pencil

Page Number	Current Selling Price	Price After the **Thousands** Digit and the **Hundred Thousands** Digit Are Exchanged	Price Difference
EXAMPLE	$659,000	$956,000	$297,000 Price Increase

STRANDS: **Number & Operations, Measurement**
SKILL: **Converting property area from acres to feet**

Acreage Conversions

Hunt for a real estate listing that gives the area of a property in *acres*. Record the page number in the first column. Copy the sentence or phrase that identifies the property size in the second column. In the third column, show the figuring that will be necessary to convert the property size from acres to square feet. Do your work on a separate piece of paper. Record the property measurement in square feet in the fourth column. Repeat these steps for six more listings. Staple your work paper to this sheet when you have completed the assignment.

Materials:

Real estate booklet, work paper, pencil

Page Number	Property Measurement in **Acres**	Figuring (show your work on a separate piece of paper)	Property Measurement in **Square Feet**
EXAMPLE	1.86 acres of private, professionally landscaped grounds	1.86 x 43,560 square feet	81,021.6 square feet

The Million-Dollar Switch

STRAND: **Number & Operations**
SKILLS: **Place value, computing differences**

Choose a house in your real estate booklet selling for over one million dollars. Record the page number and the current selling price of the house. Underline the digit in the millions place and the digit in the ten thousands place. Exchange the underlined digits and record the new price in the third column. On a separate sheet of paper, find the difference between the original price and the new price. Record the difference in the last column. Indicate whether the difference is an increase or a decrease. Repeat these steps for six more houses. Staple your work paper to this sheet when you have completed the assignment.

Materials:

Real estate booklet, work paper, pencil

Page Number	Current Selling price	Price After **Millions** Digit and **Ten Thousands** Digit Are Exchanged	Price Difference
EXAMPLE	$2,450,000	$5,420,000	$2,970,000 Price Increase

STRAND: **Number & Operations**
SKILLS: **Adding three or more addends, comparing sums**

"Sum" Great Houses for Sale

Select a house from your real estate booklet. Look at the digits that make up the house price. On a sheet of work paper, add the digits together to find the digit sum of the house. Record the page number, house price, and the number sentence you used to find the digit sum in the correct column below. Repeat these steps to fill in the whole chart. Staple your work paper to this sheet when you have completed the assignment.

Materials:

Real estate booklet, work paper, pencil, crayons

Sums Less than 12	Sums Between 12 and 21	Sums Greater than 20
page ______	EXAMPLE $519,000 (5 + 1) + 9 = 15	page ______
page ______	page ______	page ______
page ______	page ______	page ______
page ______	page ______	page ______
page ______	page ______	page ______

Use a yellow crayon to color the box containing the **lowest** sum. Color the box with the **greatest** sum orange. With a red crayon, outline each box containing an **even** sum.

Let's Have "Sum" Fun

STRANDS: **Number & Operations, Algebra**
SKILLS: **Adding three or more addends, identifying odd and even numbers, identifying multiples**

Select a house from your real estate booklet. Look at the digits that make up the price. On a sheet of work paper, add the digits together to find the digit sum of the house. Identify whether the sum is odd or even. Record the page number, house price, and the number sentence you used to find the digit sum in the correct column below. Repeat these steps to fill in the whole chart. Staple your work paper to this sheet when you have completed the assignment.

Materials:

Real estate booklet, work paper, pencil, crayons

Odd Sums	Even Sums
EXAMPLE 11 + 18 **$749,900** (7 + 4) + (9 + 9) = 29	page _____
page _____	page _____
page _____	page _____
page _____	page _____
page _____	page _____
page _____	page _____

Use a blue crayon to color the box containing the **lowest** sum. Color the box with the **greatest** sum green. Outline each box containing a sum that is a **multiple of five** with a red crayon.

Billion-Dollar Budget

STRAND: **Number & Operations**
SKILL: **Subtracting 6-digit numbers with trading**

Materials:

Real estate booklet, pencil

You've been given a billion dollars to buy homes for your family and friends! Will you choose to give away ten million-dollar mansions or share your winnings with more people by purchasing less expensive homes? Look through your real estate booklet. Choose a house for a friend or family member. List the page number in the first column. In the second column, copy a sentence or phrase from the listing. Write the name of the person you are giving the home to in the third column. Record the price of the home in the fourth column. In the last column, write the selling price again, taking care to line up the digits accurately. Subtract the price of the house from your balance of $1,000,000,000. Record the difference in the box below. This will be your starting balance when you purchase your next home. Repeat these steps until you don't have enough money to buy another home.

Page Number	Brief Description of House	Recipient of House	Selling Price	Figuring
				$1,000,000,000 –
				–
				–
				–
				–
				–
		Amount left over: (balance to bring over to next page)		

continue work on next page

(Billion-Dollar Budget—Continued from Previous Page)

Page Number	Brief Description of House	Recipient of House	Selling Price	FIGURING
Balance Brought Over from Previous Page				
				–
				–
				–
				–
				–
				–
				–
				–
				–
				–
				–
				–
				–

Spread Out!

STRAND: **Number & Operations**
SKILL: **Writing 5-digit and 6-digit numbers in expanded form**

Select a house from your real estate booklet. Record the page number in the first column and the selling price in the second column. In the third column, express the price of the house in expanded form. After completing one row, search for matching values on the game board. Color any matching boxes with a yellow crayon. Look for prices that will enable you to color in the maximum number of boxes on the game board. Repeat these steps for seven more house prices or until all squares on the game board are colored.

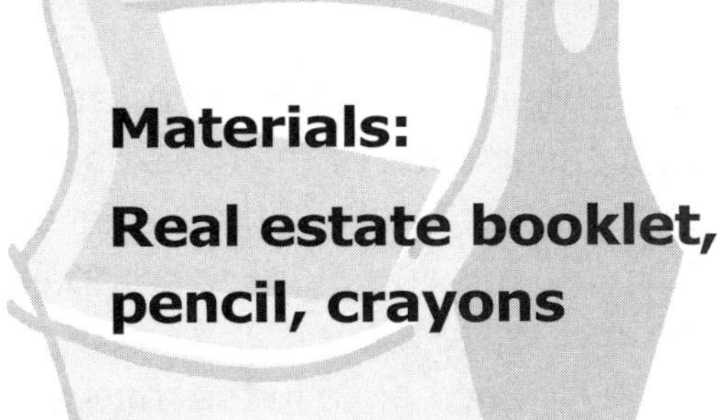

Materials:

Real estate booklet, pencil, crayons

GAME BOARD

300,000	900	1,000	800,000	50,000	5,000	60,000	400,000
9,000	500,000	1,000,000	20,000	3,000,000	700	7,000	90,000
900,000	3,000	70,000	600,000	2,000	2,000,000	500	700,000

Page Number	House Price	House Price in Expanded Form
EXAMPLE	$559,900	500,000 + 50,000 + 9,000 + 900

This House Is a Perfect Fit!

STRAND: **Number & Operations**
SKILL: **Writing the word name for 5-digit and 6-digit numbers**

Materials:

Real estate booklet, pencil

Hunt for prices in your real estate booklet that will fit perfectly in each of the rows below. When you've located a perfect match, record the page number in the first column. Fill in the missing digits to complete the price in the second column. Write the missing words to complete the word name for each price in the third column. Try your best, but you may not be able to find a price that fits into every row.

Page Number	House Price	Word Name for House Price
	$ 5__ __,000	five hundred ______________________ thousand
	$ __ __ __,500	______________________ thousand, five hundred
	$ __ __ 9,000	______________________nine thousand
	$ __ 25,000	________________ hundred twenty-five thousand
	$ __ __ __,900	______________________ thousand, nine hundred
	$1,4 __ __,000	one million, four hundred ______________ thousand
	$ __ __ 9,900	__________________nine thousand, nine hundred
	$2, __ __ __,000	two million, ______________________ thousand
	$ 3 __ __,900	three hundred ___________ thousand, nine hundred

STRAND: **Number & Operations**
SKILL: **Identifying place value relations**

How Shall I Pay for This House?

Choose a house from your real estate booklet. Write the page number in the first column and the price of the house in the second column. In the third column, record the number of hundred-dollar bills needed to purchase the house. In the fourth column, write the number of ten-dollar bills needed to purchase the house. In the last column, record the number of one-dollar bills needed to purchase the house. Repeat these steps for eight more houses.

Materials:

Real estate booklet, pencil

Page Number	House Price	...in **hundred**-dollar bills	...in **ten**-dollar bills	...in **one**-dollar bills
		Number of **$100** bills needed to purchase:	Number of **$10** needed to purchase:	Number of **$1** bills needed to purchase:
EXAMPLE	$749,900	7,499	74,990	749,900

Stop and Compare

STRAND: **Number & Operations**
SKILL: **Comparing numbers**

Locate the "Price Difference" column on the chart below. Read the expression in the first row. Look through your real estate booklet. Locate two house prices that will make the expression true. Record the house prices in the correct position on the chart. Record the page numbers where each house was found. Complete the remaining expressions.

Materials:

Real estate booklet, pencil

Page	House Price	Price Difference	House Price	Page
EXAMPLE	$995,500	is at least **$300,000** greater than	$679,900	EXAMPLE
		is between **$5,000** and **$10,000** greater than		
		>		
		is not more than **$50,000** greater than		
		is at least **$1,000,000** greater than		
		<		
		is at least **$300,000** greater than		
		is between **$10,000** and **$20,000** less than		
		is exactly **$500,000** greater than		
		>		
		is between **$100,000** and **$300,000** less than		
		is at least **$400,000** less than		

Here's My Down Payment

STRAND: **Number & Operations**
SKILLS: **Calculating percentages, comparing numbers**

Materials:
Real estate booklet, work paper, crayons, pencil

Choose a house from your real estate booklet. Write the page number in the first column and the price of the house in the second column. On a separate work paper, calculate 15%, 20%, and 25% of the selling price. Record the amounts in the appropriate columns below. Repeat this exercise with six other houses. Then follow the directions at the bottom of this paper. Staple your work paper to this sheet when you have completed the assignment.

Page Number	House Price	I'll put **15%** down...	I'll put **20%** down...	I'll put **25%** down...
EXAMPLE	$459,000	$68,850	$91,800	$114,750

Use a yellow crayon to color any box containing an **even** amount greater than $50,000 but less than $100,000. Use an orange crayon to color any box containing an **odd** amount greater than $100,000.

Putting House Prices to the Test

STRAND: **Number & Operations**
SKILLS: **Computation, place value, rounding numbers**

Choose two real estate listings from your booklet. Each listing must have a different price, and each listing must be less than $1,000,000. Cut out the listings. Glue one listing in the box labeled *Listing A* and glue the other listing in the box labeled *Listing B*. Follow the directions in the boxes below. Be sure to check your work carefully.

Materials:

Real estate booklet, scissors, glue stick, pencil

LISTING A	**LISTING B**
1A. Find the sum of the digits in the price.	1B. Find the sum of the digits in the price.
2A. Round the price to the nearest ten thousand.	2B. Round the price to the nearest ten thousand.
3A. Round the price to the nearest hundred thousand.	3B. Round the price to the nearest hundred thousand.

4. Find the sum of the two prices.

5. Find the difference between the two prices.

6A. Write the word name for the price.	6B. Write the word name for the price.
7A. Write the price in expanded form.	7B. Write the price in expanded form.

The Price Is Right!

STRAND: **Number & Operations**
SKILLS: **Computation, place value, rounding numbers**

Choose a listing from your real estate booklet. The price of the house must be less than $1,000,000. Cut out the listing. Glue the listing and the accompanying photograph to the back of this sheet. Use the price of the house to complete this paper. Be sure to check your work carefully.

Materials:

Real estate booklet, scissors, glue stick, pencil

What is the price of the house?	Write the word name for the price.
Write the price in expanded form.	
What is the value of the digit in the ten thousands place?	What is the value of the digit in the hundred thousands place?
Round the price to the nearest ten thousand.	Round the price to the nearest hundred thousand.
How many $100 bills are needed to purchase the house?	How many $10 bills are needed to purchase the house?
Find the sum of the digits that make up the price.	Increase the price of the house by $15,000.
Switch the digit in the hundreds place with the digit in the hundred thousands place. Write the new price.	What is the difference between the two the

Decimal Delight

STRAND: **Number & Operations**
SKILL: **Reading and writing fractions and decimals**

Look for examples of fractions and decimals in your real estate booklet. Each time you find an example, list the page number in the first column and write the sentence or phrase that includes the fraction or decimal in the second column. In the third column, write the fraction or decimal in words. Finally, if you located a fraction, rewrite it as a decimal. If you found a decimal, rewrite it as a fraction. Continue to hunt for other examples of fractions and decimals.

Materials:

Real estate booklet, pencil

Page Number	Fraction or decimal found in the real estate advertisement	Fraction or decimal written in words	Fraction rewritten as a decimal or decimal rewritten as a fraction
EXAMPLE	Stunning interior, 3/4 bedrooms	three fourths	.75

Let's Line Up!

STRAND: **Number & Operations**
SKILLS: **Ordering numbers, identifying multiples of 5, identifying odd numbers**

Find a real estate listing that includes a variety of whole numbers, decimals, fractions, and mixed numbers. Record the page number. Cut out the listing and glue it in the space provided. Highlight the whole numbers, decimals, fractions, and mixed numbers that appear in the listing. Order all of the numbers from least to greatest in the space provided. Underline each odd whole number. Mark an "X" above each whole number that is a multiple of five. Repeat these steps to complete the page.

Materials:

Real estate booklet, scissors, glue stick, highlighter, pencil

EXAMPLE	**CONVENIENT LOCATION!** Sit on your deck overlooking **420** ft. of direct waterfront on fully recreational pond. This **12** room, **5** bedroom, **3 1/2** bath custom Colonial with **2** car garage has it all. **21'** kitchen, library with hardwoods. Over **5,400** square feet of living space. Set on a **.75** acre nicely treed lot. **$1,279,000**
	X above 5, 420 (X above 3,400), X above $1,279,500 .75, 2, 3 1/2, <u>5</u>, 12, <u>21</u>, 420, 3,400, $1,279,500
Page ____	
Page ____	
Page ____	
Page ____	

Symmetrical Search

STRAND: **Geometry**
SKILLS: **Identifying lines of symmetry, recognizing and describing symmetrical figures**

Look through your real estate book for photographs of houses, or parts of houses, that are symmetrical. You may choose to cut out the garage, the second floor, or the first floor of a house. Perhaps you'll cut out a shed or barn located on the property. In rare cases, the whole house will be symmetrical. Glue the example of symmetry in the first box below. Use a pen or marker to make a dashed *line of symmetry* on the example. In the space provided, answer the question, "How can you prove that this figure is symmetrical?" Repeat these steps to complete the page.

Materials:

Real estate booklet, scissors, glue stick, marker, pencil

Symmetrical Example #1	Symmetrical Example #2
How can you prove that this figure is symmetrical?	How can you prove that this figure is symmetrical?

STRAND: **Geometry**
SKILL: **Identifying similar and congruent figures**

Similar or Congruent?

Look at the photographs of houses in your real estate booklet. Examine the size and shape of the windows, shutters, and doors on the houses. Hunt for examples of congruent and similar figures. Cut out a photograph from one of your favorite listings. Glue the photograph in the first box. In the space provided, describe the figures on the house using the vocabulary words *congruent* and *similar.* Repeat these steps to complete the page.

Materials:

Real estate booklet, scissors, glue stick, pencil

Similar and Congruent Example #1	Similar and Congruent Example #2
Describe the figures on the house using the vocabulary words **congruent** and **similar**.	Describe the figures on the house using the vocabulary words **congruent** and **similar**.

Perimeter Priority

STRAND: **Number & Operations, Measurement**
SKILL: **Finding the perimeter of a rectangle**

Locate a rectangular advertisement in your real estate booklet. Cut out the advertisement and glue it to the construction paper. Label it with the letter A. Measure its length and width in centimeters. Record the measurements on the chart below. Add the length of the four sides to find the perimeter. Record the perimeter in the last column. Repeat these steps four more times. Then answer the question at the bottom of this paper. Staple the construction paper to this sheet.

Materials:

Real estate booklet, scissors, glue stick, 12 x 18-inch sheet of construction paper, metric ruler, pencil

Rectangle	Length	Width	Figuring	Perimeter
EXAMPLE	6 centimeters	4.5 centimeters	6 + 6 + 4.5 + 4.5 = 21	21 centimeters
A				
B				
C				
D				
E				

In centimeters, what is the difference between the perimeter of your smallest rectangle and the perimeter of your largest rectangle? Show your work.

Area Accuracy

STRAND: **Number & Operations, Measurement**
SKILL: **Finding the area of a rectangle**

Locate a rectangular advertisement in your real estate booklet. Cut out the advertisement and glue it to the construction paper. Label it with the letter A. Measure its length and width in centimeters. Record the measurements on the chart below. Multiply the length and width to find the area. Record the area in the last column. Repeat these steps four times. Then answer the question at the bottom of this paper. Staple the construction paper to this sheet.

Materials:

Real estate booklet, scissors, glue stick, 12 x 18-inch sheet of construction paper, metric ruler, pencil

Rectangle	Length	Width	Figuring	Area
EXAMPLE	3 centimeters	5 centimeters	3 x 5 = 15	15 square centimeters
A				
B				
C				
D				
E				

If each of your rectangles were placed together to form one large figure, what would the area of the figure be? Show your work.

The Right Angle

STRAND: **Geometry**
SKILLS: **Identifying right angles, acute angles, obtuse angles**

Look at the houses in your real estate booklet. Locate a house that has at least one example of a right angle, an acute angle, *and* an obtuse angle. Cut out the house and glue it in the middle of Box A. Use arrows to point out one example of each angle. Write the name of the angle next to each arrow. Write a sentence to explain your label, such as, "This window frame forms a right angle." Repeat these steps for Box B.

Materials:

Real estate booklet, scissors, glue stick, pencil

BOX A

BOX B

House Prices Tally Chart

STRANDS: **Number & Operations, Data Analysis & Probability**
SKILLS: **Creating and interpreting a tally chart**

Materials:

Real estate booklet, pencil

Look at the price of the first house listed in your real estate booklet. Decide which category the house fits into on the tally chart below. Place a tally mark in the appropriate row. Place a check mark next to the house in your booklet to indicate that you've included that house on your tally chart. Repeat these steps for at least 25 houses. When you have completed your tally chart, count the tally marks in each row. Write the total number of tally marks in the last column.

House Price	Tally Marks	Total
Less than **$450,000**		
From **$450,000** to **$600,000**		
From **$601,000** to **$750,000**		
From **$751,000** to **$900,000**		
From **$901,000** to **$1,050,000**		
Greater than **$1,050,000**		

On the back of this paper, write five true statements about your completed tally chart. Use three or more of the following words or phrases: **sum**, **difference**, **greater than**, **less than**, **equal**, **odd**, **even**, **more than half**, **less than half**, **least**, and **greatest**.

House Prices Pictograph

STRANDS: **Number & Operations, Data Analysis & Probability**
SKILL: **Creating and interpreting a pictograph**

Draw the outline of a pentagon-shaped house symbol in the key at the bottom of this chart. The symbol must be symmetrical, so do not add a chimney. One house symbol will represent the data from two listings. Look at the total number of tally marks beside each category in your House Prices Tally Chart. Figure out how many house symbols you will need to represent the totals in each of the categories below. Draw in the symbols. If necessary, use half of a symbol to represent one listing.

Materials:

Real estate booklet, your completed House Prices Tally Chart, pencil

House Price	Symbols
Less than **$450,000**	
From **$450,000** to **$600,000**	
From **$600,001** to **$750,000**	
From **$750,001** to **$900,000**	
From **$900,001** to **$1,050,000**	
Greater than **$1,050,000**	
KEY:	= Two houses

On the back of this paper, write five true statements about your completed pictograph. Use three or more of the following words or phrases: **sum**, **difference**, **greater than**, **less than**, **equal**, **odd**, **even**, **more than half**, **less than half**, **least**, and **greatest**.

House Prices Bar Graph

STRANDS: **Number & Operations, Data Analysis & Probability**
SKILLS: **Creating and interpreting a bar graph**

Materials:

Real estate booklet, crayons, pencil

Choose a scale for your bar graph. Fill in the scale in the first column below. Locate a house in your real estate booklet. Read the price. Decide which column the house fits into. Based on your graph's scale, color in the appropriate number of spaces in the correct column to represent that house. Place a check mark next to the house in your booklet to indicate that you have already added that house to your bar graph. Repeat these steps for at least 25 houses. Add a title and labels to your graph.

Title: ______________________________

	Less than **$450,000**	From **$450,000** to **$600,000**	From **$600,001** to **$750,000**	From **$750,001** to **$900,000**	From **$900,001** to **$1,050,000**	Greater than **$1,050,000**

On the back of this paper, write five true statements about your completed bar graph. Use three or more of the following words or phrases: **sum**, **difference**, **greater than**, **less than**, **equal**, **odd**, **even**, **more than half**, **less than half**, **least**, and **greatest**.

STRANDS: **Number & Operations, Data Analysis & Probability**
SKILL: **Creating and interpreting a tally chart**

Counting Bedrooms Tally Chart

Look at the first house listed in your real estate booklet. Highlight the number of bedrooms identified. Decide which category the house fits into on the chart below. Place a tally mark in the appropriate row. If a listing does not identify the number of bedrooms, place a tally mark in the last category: Number of Bedrooms Not Identified. By highlighting the number of bedrooms, you'll keep track of the listings you've already used. Repeat these steps for at least 25 houses. When you have completed your chart, count the tally marks in each row. Write the total number of tally marks in the last column.

Materials:

Real estate booklet, highlighter, pencil

Number of Bedrooms Identified in Listing	Tally Marks	Total
1 Bedroom		
2 Bedrooms		
3 Bedrooms		
4 Bedrooms		
5 Bedrooms		
6 or More Bedrooms		
Number of Bedrooms Not Identified		

On the back of this paper, write five true statements about your completed tally chart. Include the following vocabulary words or phrases: **sum**, **difference**, **greater than**, **less than**, **equal**, **odd**, **even**, **more than half**, **less than half**, **least**, and **greatest** in your statements.

STRANDS: **Number & Operations, Data Analysis & Probability**
SKILL: **Creating and interpreting a pictograph**

Counting Bedrooms Pictograph

Draw a bed symbol in the key at the bottom of this chart. The symbol must be symmetrical. One bed will represent the data from two listings. Look at the total number of tally marks beside each category in your Counting Bedrooms Tally Chart. Figure out how many bed symbols you will need to represent the totals in each of the categories below. Draw in the symbols. If necessary, use half of a symbol to represent one listing.

Materials:

Real estate booklet, completed Counting Bedrooms Tally Chart, pencil

Number of Bedrooms	Symbols
1 Bedroom	
2 Bedrooms	
3 Bedrooms	
4 Bedrooms	
5 Bedrooms	
6 or More Bedrooms	
Number of Bedrooms Not Identified	

KEY: = Data from two listings

On the back of this paper, write five true statements about your completed pictograph. Include the following vocabulary words or phrases: **sum**, **difference**, **greater than**, **less than**, **equal**, **odd**, **even**, **more than half**, **less than half**, **least**, and **greatest** in your statements.

STRANDS: **Number & Operations, Data Analysis & Probability**
SKILL: **Creating and interpreting a bar graph**

Counting Bedrooms Bar Graph

Choose a scale for your bar graph. Fill in the scale on the chart below. Locate a house in your real estate booklet. Decide which column the data fits into. Based on your graph's scale, color in the appropriate number of spaces in the correct column to represent the data. If a listing does not identify the number of bedrooms, use the last category: Number of Bedrooms Not Identified. Place a check mark next to the house in your booklet to keep track of the listings you've already used. Repeat these steps for at least 12 houses. Add a title and labels to your graph.

Materials:

Real estate booklet, crayons, pencil

Title: ______________________________

	1 Bedroom	2 Bedrooms	3 Bedrooms	4 Bedrooms	5 Bedrooms	6 or More Bedrooms	Number of Bedrooms Not Identified

On the back of this paper, write five true statements about your completed bar graph. Include the following vocabulary words or phrases: **sum**, **difference**, **greater than**, **less than**, **equal**, **odd**, **even**, **more than half**, **less than half**, **least**, and **greatest** in your statements.

Living Space Tally Chart

STRANDS: **Number & Operations, Data Analysis & Probability**
SKILL: **Creating and interpreting a tally chart**

Materials:

Real estate booklet, highlighter, pencil

Look in your real estate booklet for a listing that identifies the area of living space in square feet. Highlight the area. Look at the tally chart below. Decide which category the house fits into. Place a tally mark in the appropriate row. Skip over any listings that do not identify the area of living space. By highlighting the area, you'll keep track of the listings you've already used. Repeat these steps for at least 25 listings. When you have completed your tally chart, count the tally marks for each category. Write the total number of tally marks in the last column.

Area of Living Space Identified in Listing	Tally Marks	Total
Less Than **2,000** Square Feet		
From **2,000** to **3,000** Square Feet		
From **3,001** to **4,000** Square Feet		
From **4,001** to **5,000** Square Feet		
From **5,001** to **6,000** Square Feet		
Greater Than **6,000** Square Feet		

On the back of this paper, write five true statements about your completed tally chart. Include the following vocabulary words or phrases: **sum**, **difference**, **greater than**, **less than**, **equal**, **odd**, **even**, **more than half**, **less than half**, **least**, and **greatest** in your statements.

Living Space Pictograph

STRANDS: **Number & Operations, Data Analysis & Probability**
SKILL: **Creating and interpreting a pictograph**

Draw a capital letter A in the key at the bottom of this chart. The letter A is the symbol that will represent the data from two listings. Look at the total number of tally marks beside each category in your Living Space Tally Chart. Figure out how many letter A symbols you will need to represent the totals in each of the categories below. Draw in the symbols. If necessary, use half of a symbol to represent one listing.

Materials:

Real estate booklet, completed Living Space Tally Chart, pencil

Area of Living Space Identified in Listing	Symbols
Less Than **2,000** Square Feet	
From **2,000** to **3,000** Square Feet	
From **3,001** to **4,000** Square Feet	
From **4,001** to **5,000** Square Feet	
From **5,001** to **6,000** Square Feet	
Greater Than **6,000** Square Feet	
Key:	= Data from two listings

On the back of this paper, write five true statements about your completed pictograph. Include the following vocabulary words or phrases: **sum**, **difference**, **greater than**, **less than**, **equal**, **odd**, **even**, **more than half**, **less than half**, **least**, and **greatest** in your statements.

Living Space Bar Graph

STRANDS: **Number & Operations, Data Analysis & Probability**
SKILL: **Creating and interpreting a bar graph**

Materials:

Real estate booklet, crayons, pencil

Choose a scale for your bar graph. Fill in the scale in the first column below. Locate a house in your real estate booklet that identifies the area of living space in square feet. Decide which column the data fits into. Based on your graph's scale, color in the appropriate number of spaces in the correct column to represent the data. Skip any listings that do not identify the area of living space. Place a check mark next to the house in your booklet to keep track of the listings you've already used. Repeat these steps for at least 25 houses. Add a title and labels to your graph.

Title: ____________________________

	Less Than **2,000** Square Feet	From **2,000** to **3,000** Square Feet	From **3,001** to **4,000** Square Feet	From **4,001** to **5,000** Square Feet	From **5,001** to **6,000** Square Feet	Greater Than **6,000** Square Feet

On the back of this paper, write five true statements about your completed bar graph. Include the following vocabulary words or phrases: **sum**, **difference**, **greater than**, **less than**, **equal**, **odd**, **even**, **more than half**, **less than half**, **least**, and **greatest** in your statements.

Acreage Tally Chart

STRANDS: **Number & Operations, Data Analysis & Probability**
SKILL: **Creating and interpreting a tally chart**

Materials:

Real estate booklet, highlighter, pencil

Look in your real estate booklet for a listing that identifies the size, in acres, of the piece of land a house sits on. Highlight the land size. Look at the tally chart below. Decide which category the house fits into. Place a tally mark in the appropriate row. Skip listings that do not identify the acreage. The highlighted acreage will indicate that you have already added the data from that listing to your tally chart. Repeat these steps for at least 25 listings. When you have completed your tally chart, count the tally marks for each category. Write the total number of tally marks in the last column.

Acreage of Land the House Sits On	Tally Marks	Total
Less than **.5** acres		
From **.5** to **1.5** acres		
From **1.6** to **2.5** acres		
From **2.6** to **3.5** acres		
From **3.6** to **4.5** acres		
From **4.6** to **5.5** acres		
Greater than **5.6** acres		

On the back of this paper, write five true statements about your completed tally chart. Include the following vocabulary words or phrases: **sum**, **difference**, **greater than**, **less than**, **equal**, **odd**, **even**, **more than half**, **less than half**, **least**, and **greatest** in your statements.

Acreage Pictograph

Draw a square symbol in the key at the bottom of this chart. One square will represent two listings. Look at the total number of tally marks beside each category in your Acreage Tally Chart. Figure out how many square symbols you will need to represent the totals in each of the categories below. Draw in the symbols. When half of a symbol is needed to represent one listing, use a square cut on the diagonal to represent the data.

Materials:

Real estate booklet, completed Acreage Tally Chart, pencil

ACREAGE OF LAND THE HOUSE SITS ON	SYMBOLS
Less than **.5** acres	
From **.5** to **1.5** acres	
From **1.6** to **2.5** acres	
From **2.6** to **3.5** acres	
From **3.6** to **4.5** acres	
From **4.6** to **5.5** acres	
Greater than **5.6** acres	
KEY:	= Data from two listings

On the back of this paper, write five true statements about your completed tally chart. Include the following vocabulary words or phrases: **sum**, **difference**, **greater than**, **less than**, **equal**, **odd**, **even**, **more than half**, **less than half**, **least**, and **greatest** in your statements.

STRANDS: **Number & Operations, Data Analysis & Probability**
SKILL: **Creating and interpreting a bar graph**

Acreage Bar Graph

Choose a scale for your bar graph. Fill in the scale in the first column below. Locate a listing in your real estate booklet that identifies the size, in acres, of the piece of land a house sits on. Decide which column the data fits into. Based on your graph's scale, color in the appropriate number of spaces in the correct column to represent the data. Skip listings that do not identify the acreage. Place a check mark next to the house in your booklet to indicate that you have already added the data from that house to your graph. Repeat these steps for at least 25 listings. Add a title and labels to your graph.

Materials:

Real estate booklet, crayons, pencil

Title: ______________________________

	Less than .5 acres	From .5 to 1.5 acres	From 1.6 to 2.5 acres	From 2.6 to 3.5 acre	From 3.6 to 4.5 acres	From 4.6 to 5.5 acres	Greater than 5.6 acres

On the back of this paper, write five true statements about your completed bar graph. Include the following vocabulary words or phrases: **sum**, **difference**, **greater than**, **less than**, **equal**, **odd**, **even**, **more than half**, **less than half**, **least**, and **greatest** in your statements.

BUILD A BOOK

Students use the listings found in their real estate booklets to create their own math concept books.

Build a Book projects offer students a creative way to demonstrate mastery of their mathematics skills. These activities work wonderfully when presented to the whole class. I find it's best to assign a Build a Book project after your class has had much practice with the skills involved in the particular activity. (See the chart on page 78 for quick reference.)

Set aside a one-hour block of time to kick off a Build a Book project. Once your students complete the first couple of pages, they'll be ready to work independently. How much time your students will work on a project and the length of each book is up to you. You'll find a minimum number of suggested pages listed for each activity.

MATERIALS: In addition to a real estate booklet, students will need scissors, glue, pencils, crayons or markers, and 12 x 18-inch construction paper. Teacher preparation is minimal. After students complete all of their pages, you'll need a stapler, or a hole-punch and fasteners, to help students bind their books.

FOLLOW-UP: For additional mathematics practice, I require that each book is reviewed for accuracy by two peers. Working independently, each student editor must check all computation using a calculator and verify the accuracy of every statement. I always have plenty of sticky notes on hand for student editors to mark any errors they find. When the editors are finished, they add their names to a "This Book Was Edited By..." sticker attached to the back cover. Then, I conference with the individual authors and assist them in correcting any errors that have been identified.

Students love sharing their finished books with other classes at their grade level or reading them to students in younger grades! You may want to store the books in a basket in your classroom library and allow students to borrow them overnight so they can be enjoyed at home. These books are sure to become very popular, so you may want to laminate them!

TEACHER TIP: All students, regardless of their ability level, will experience success with the Build a Book projects. Every activity includes a list of modifications for students with special needs and a list of challenges for high-achievers. Use one or more of these suggestions as appropriate for your students.

Skills Required for Each Build a Book Project

Title:	Skills Reinforced:
The Houses That Jack Bought	• adding 6-digit and 7-digit numbers • writing ordinal numbers • writing money amounts through the millions
Counting on Real Estate	• ordering whole numbers, decimals, fractions, and mixed numerals from least to greatest • writing the word name for whole numbers, decimals, and fractions
Order in the House	• writing the word name for ordinal number abbreviations • writing ordinal number abbreviations for ordinal number word names • arranging ordinal numbers from least to greatest
It All Adds Up	• writing addition story problems for items found in photographs • solving simple addition story problems • identifying the least sum and the greatest sum
What's the Difference?	• writing subtraction story problems for items found in photographs • solving simple subtraction story problems • identifying the least difference and the greatest difference
Unlikely Listings	• creating simple and complex expressions for whole numbers, decimals, fractions, and mixed numerals • converting feet to inches

STRAND: **Number & Operations**
SKILLS: **Computation, using ordinal numbers, writing numbers through the millions**

The Houses That Jack Bought

Jack isn't building houses any more—he's *buying* them! Your class will enjoy putting their addition skills to the test as they each create a book inspired by the popular children's rhyme, *The House That Jack Built*. In addition to assessing the computational skills of your students, you'll also have an opportunity to evaluate their knowledge of ordinal numbers and their ability to write numbers through the millions.

Directions:

1. To kick off this activity, read some of the many book versions of *The House That Jack Built*. You may want to begin with some traditional versions and end with my favorite, the absolutely hilarious *This Is the House That Jack Built*, by Simms Taback. This entertaining, up-to-date tale has many surprises that will have your students laughing and eager to write their own books.

2. Invite students to look through their real estate booklets for houses that Jack would like to buy. Ask each student to cut out one listing and its accompanying photograph. Each student glues his cut-out pieces onto construction paper as shown in the sample below. Under the photograph, the student writes:

   This is the 1st house that Jack bought. It cost ______________.

   But Jack wasn't done shopping.

Each student includes the price of his house as shown. This is the first page of the student's book.

3. For page two, each student cuts and glues another listing and photograph to his construction paper. To the right of the photograph, the student adds the price of the first house to the price of the second house in working form. Under the photograph, the student writes:

This is the 2nd house that Jack bought. It cost ______. In all, he had spent __________________.

But Jack wasn't done shopping.

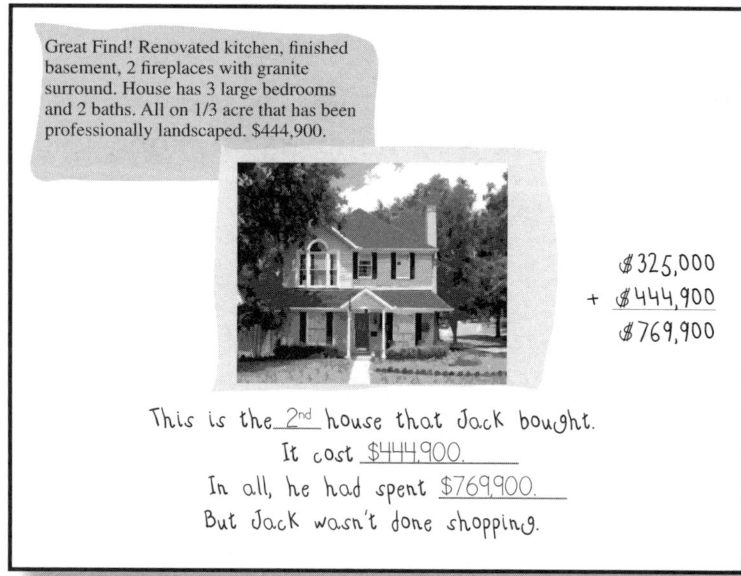

Students include the sum of their first and second houses' prices as shown. This is page two of the book.

4. Instruct the class to continue creating additional pages for their books. Remind students to add Jack's previous spending total to the price of each new house. The text pattern will remain the same on each new page. The only items that change are the ordinal number, the cost of the last house purchased, and the total sum of the houses that Jack has purchased.

5. For the final page, have students change the last line of text to:

   Jack was finally done shopping.

On each page, students find the sum total of the homes purchased—it's great computation practice!

Minimum Number of Pages:

Choose a page length based on the time available and the ability level of your students. Books of at least five pages will allow you to assess your students' ability to add five-digit and six-digit numbers.

Cover Directions:

Have students create neat, colorful covers that include the title, the author's full name, and a picture of Jack purchasing one of his homes.

MODIFICATIONS FOR STUDENTS WITH SPECIAL NEEDS:

- Use graph paper for the book pages so students may line up addition problems with ease.
- Allow students to use calculators for computation.
- Copy the text of the story on strips of paper. Include blanks for students to fill in the ordinal numbers, the house prices, and the running total. Allow students to glue these strips onto their book pages.

CHALLENGES FOR HIGH-ACHIEVING STUDENTS:

- Next to prices written in standard form, challenge students to write the word name for each price.
- Under the line that tells the cost of the house, challenge students to add the line, "The cost of this house was ____________ more/less than the last house that he bought." The last two lines of the text will remain the same.
- Challenge students to have Jack spend exactly $2,000,000 dollars.

Counting on Real Estate

Traditional counting books usually feature whole numbers in counting order. *Counting on Real Estate* includes whole numbers, decimals, fractions, and mixed numbers. I like to conference with students as they write these unique counting books. Keep a checklist handy as you circulate from one author to the next. Record students' ability to read and compare whole numbers, fractions, and decimals. Later, as you review the completed books, you'll be able to assess how well students are able to order whole numbers, fractions, and decimals from least to greatest.

Directions:

1. Kick off this activity by sharing a collection of counting books with the class. Try to include books that move away from the traditional 1-2-3 counting pattern. I'm a big fan of Jerry Pallotta's clever counting books. In *Underwater Counting: Even Numbers,* he counts by twos to 50 using even numbers, while in *Ocean Counting: Odd Numbers*, he counts by twos using only odd numbers. In *Reese's Pieces Count by Fives,* he counts, of course, by fives, and in *Count to a Million*, he counts through the decimal system.

2. Give students time to look through their real estate booklets for listings that include whole numbers, decimals, fractions, or mixed numbers. Have each student cut out one listing and its accompanying photograph. Students glue their cut-out pieces to their construction paper as shown in the sample below.

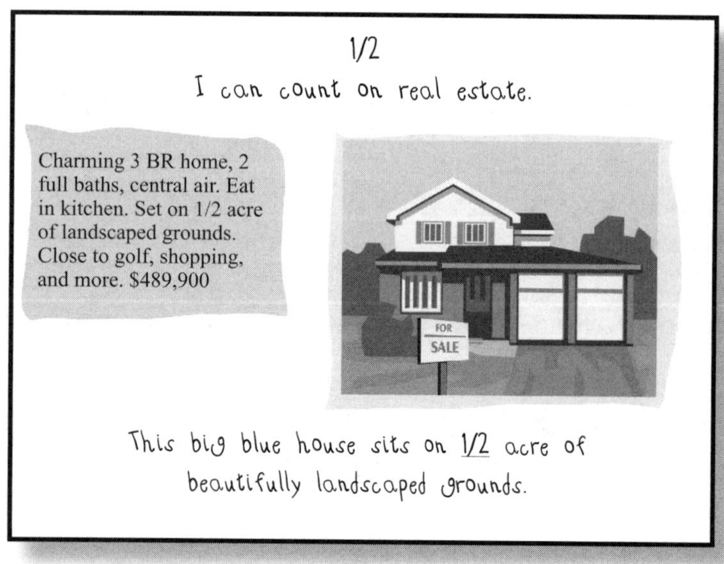

3. Since most real estate listings include several numbers, instruct each student to select and highlight the one fraction, decimal, or whole number she wants to feature on that page. The student writes her featured number on the top of her construction paper, as shown in the sample below, using a dark crayon or marker. Under the featured number, the student writes:

   I can count on real estate.

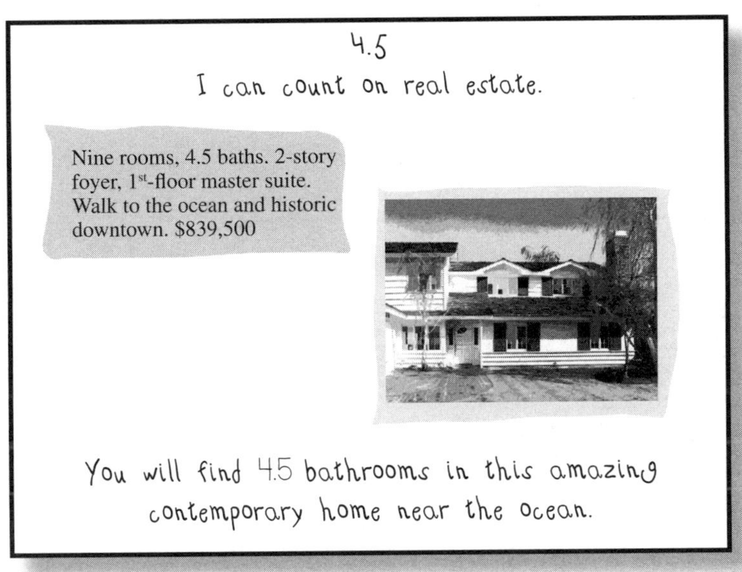

4. Next, ask each student to write a sentence on the bottom of her paper that includes her featured number. The sentence should use some of the information found in the listing, but it must be written in the student's own words.

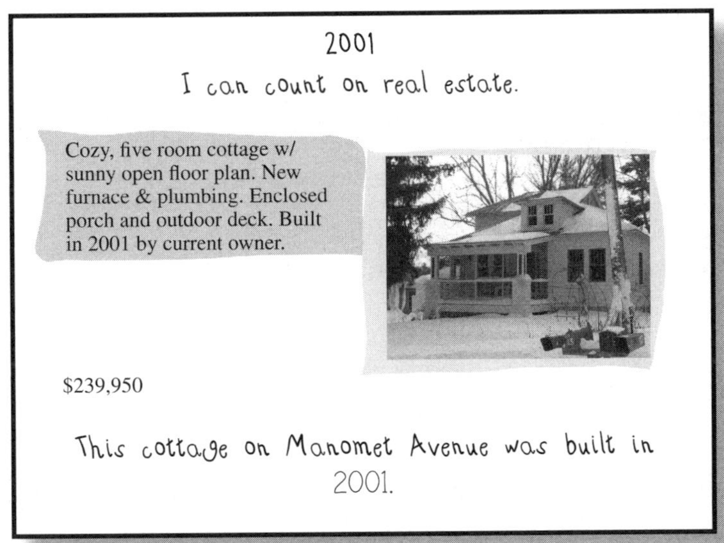

5. When students complete all of their pages, have them order the pages from least to greatest.

Minimum Number of Pages:

Choose a page length based on the time available and the ability level of your students. Books of eight to ten pages will allow you to assess your students' ability to order whole numbers, decimals, fractions, and mixed numbers from least to greatest.

Cover Directions:

Invite students to create neat, colorful covers that include the title and the author's full name. The covers may be decorated with all of the whole numbers, decimals, fractions, and mixed numbers featured in the book.

MODIFICATIONS FOR STUDENTS WITH SPECIAL NEEDS:

- Ask students to include only whole numbers in the book.
- Limit the range of numbers that will be included in the book.
- Allow each student to copy the sentence or phrase from the listing that includes her number rather than writing the text on her own.

CHALLENGES FOR HIGH-ACHIEVING STUDENTS:

- Ask students to write the word name next to each whole number, decimal, fraction, or mixed number.
- Invite students to convert each decimal to a fraction and each fraction to a decimal. The conversion may be written in parentheses, to the right of the original decimal or fraction.

Order in the House

Ordinal numbers are found in many real estate listings. They are used to identify floors, bedrooms, and other features of a home and its grounds. Sometimes ordinal numbers are even used to identify the location of the house as it relates to the holes on a golf course! Students will welcome the opportunity to hunt for ordinal numbers and ordinal number words as they create a book called *Order in the House*. Completed books will allow you to assess your students' ability to read, write, and arrange ordinal numbers.

Directions:

1. Kick off this activity with a quick review of ordinal numbers. You may want to begin by discussing the dates on the calendar, the floors in a high-rise building, the grades between kindergarten and college, or the holes on a golf course. Depending on the time of year, you may choose to sing "The Twelve Days of Christmas" or read *The Thirteen Days of Halloween* by Carol Greene.
2. Invite the class to look through their real estate booklets for listings that include ordinal numbers. These numbers may be a little difficult to find so urge students to be patient. Have each student cut out one listing and its accompanying photograph. Instruct students to glue their cut-out pieces to their sheet of construction paper as shown in the sample below.

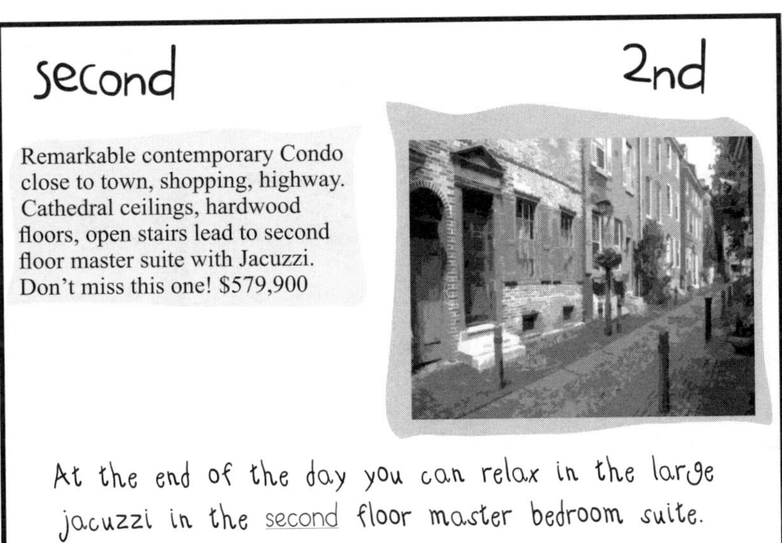

3. Ask each student to use a dark crayon or marker to write the ordinal number word featured in his listing in the top left corner of his paper. Have the student write the abbreviation for that ordinal number in the top right corner of his paper.

4. Next, each student writes a sentence on the bottom of his paper that includes his ordinal number. The sentence should use some of the information found in the listing, but it must be written in the student's own words.

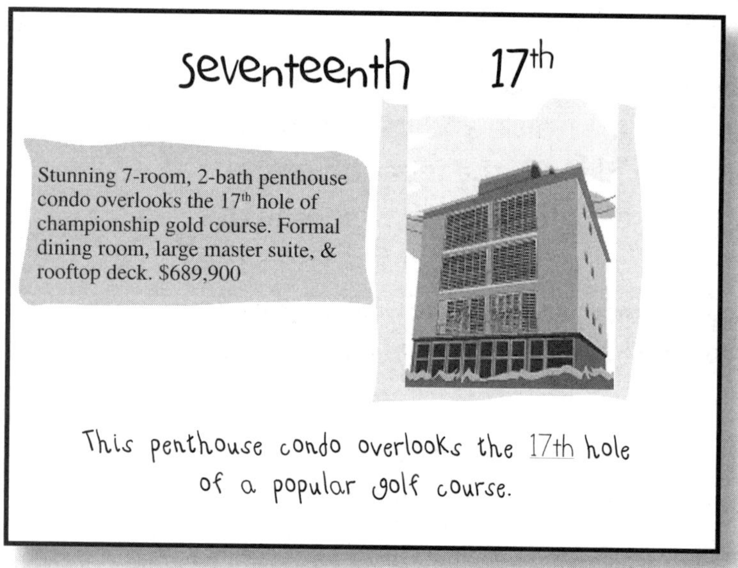

5. When students complete all of their pages, have them order the pages from least to greatest.

Minimum Number of Pages:

Choose a page length based on the time available and the ability level of your students. Books of five or more pages will allow you to accurately assess your students' ability to order whole numbers, fractions, and decimals from least to greatest.

Cover Directions:

Invite students to create neat, colorful covers that include the title and the author's full name. The covers may be decorated with all of the ordinal number abbreviations included in the book.

MODIFICATIONS FOR STUDENTS WITH SPECIAL NEEDS:

- Locate listings that include ordinal numbers and ordinal number words in advance.
- Use a highlighter to draw attention to the ordinal numbers or ordinal number words.
- Allow each student to copy the sentence or phrase from the listing that includes his ordinal number or ordinal number word rather than writing the text on his own.

CHALLENGES FOR HIGH-ACHIEVING STUDENTS:

- Invite students to research how their ordinal numbers may be written as Roman numerals.
- Have each student construct a two-column chart on the final page of his book. The chart should include each ordinal number word and abbreviation featured in his book.

Sample Ordinal Number Chart:

Ordinal Number Abbreviation	**Ordinal Number Word**
1st	first
2nd	second
3rd	third
10th	tenth
13th	thirteenth

It All Adds Up

Shutters, doors, chimneys, and steps! Shrubs, trees, flags, and planters! Mail boxes, windows, columns, and lamp posts! Once students start looking at the photographs in their real estate booklets for items to add, the possibilities will seemingly leap from the pages. The choices are endless. In *It All Adds Up*, students write addition stories. The completed books allow you to assess each student's ability to create addition number stories, write number sentences, and solve basic addition facts.

Directions:

1. To kick off this activity, pass out magnifying glasses to the students and ask them to look closely at the photographs in their real estate booklets. Invite the students to study the photographs and tell you what they see on and around the homes. List these features on the chalkboard or a piece of chart paper. The list usually includes windows, lights, shutters, steps, chimneys, doors, shrubs, trees, flags, and planters.
2. Ask each student to cut out one listing and its accompanying photograph. Have students glue their cut-out pieces to their construction paper as shown in the sample below.
3. Instruct each student to select two different features shown on the house or in the yard. Features might include windows, lights, shutters, steps, chimneys, doors, shrubs, trees, flags, or planters.

On the front of the house, there are 13 windows, and 20 shutters. What is the total number of windows and shutters that may be seen on the front of the house?

Beautiful West Plymouth home in sought after neighborhood. Updated kitchen with granite counters. Large family room w/ skylight. Central vac, sprinkler system. $673,500

13 + 20 = 33 windows and shutters

4. Next, ask each student to write a short addition story problem that incorporates the two features she selected above her listing. Remind students that every story problem includes a question!
5. Below the listing, have each student write the number sentence needed to solve her story problem. The sum and the appropriate label must be included.

6. When students complete all of their pages, have them order the pages from the number sentence with the greatest sum to the number sentence having the lowest sum.

Minimum Number of Pages:

Choose a page length based on the time available and the ability level of your students. Books of five or more pages will allow you to accurately assess your students' ability to write addition story problems and solve addition number sentences.

Cover Directions:

Invite students to create neat, colorful covers that include the title and the author's full name. The covers may be decorated with the number sentence having the greatest sum and the number sentence having the lowest sum.

MODIFICATIONS FOR STUDENTS WITH SPECIAL NEEDS:

- Cut out the listings and photographs in advance.
- Provide students with photographs of houses that have windows with shutters and give them copies of the story problem form below. Students glue copies of the story form above each listing and fill in the correct numbers. On the bottom of the paper, each student writes the number sentence needed to solve the story problem.

The front of the house has _____ windows. There are also ______ shutters on the front of the house. How many windows and shutters in all?

CHALLENGES FOR HIGH-ACHIEVING STUDENTS:

- Invite students to write more complex story problems that require a number sentence with three or four addends. Challenge students to use parentheses in their number sentences.

What is the total number of windows that have been installed on the front of the house if there is 1 large window to the left of the door, 4 small windows to the left of the door, and 3 large windows above the door?

This cute 3 bedroom home has so much charm. Fireplace in living room and hardwood floors. Laundry area off kitchen. Landscaped backyard. $236,500

1 + (4 + 3) = 8 windows

- Invite each student to place two listings and two photographs on each page and ask her to create story problems using the characteristics of both houses.

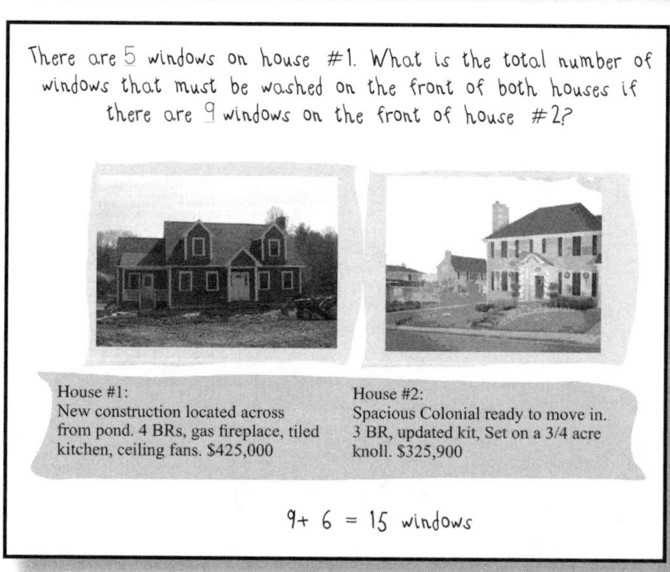

STRAND: **Number & Operations**
SKILLS: **Comparing amounts, creating and solving subtraction word problems, ordering differences**

What's the Difference?

As buyers hunt for their dream home, they're constantly comparing features and looking carefully at every detail. Nothing is overlooked. In *What's the Difference?*, your students scrutinize the photographs in their real estate booklets to create a book of subtraction stories. As you review the completed books, you'll have the opportunity to assess your students' ability to create subtraction number stories, write number sentences, and solve basic subtraction facts.

Directions:

1. To kick off this activity, display the words *more* and *fewer* on large cards and two color photographs of homes cut from a real estate booklet. Invite students to compare the features of the homes using the words *more* and *fewer*. For example, a student might say, "The red house has two *fewer* windows than the white house," or "There are five *more* trees in the yard of the red house than in the yard of the white house." After sharing several examples, the students will be ready to begin writing subtraction stories of their own.
2. Invite students to look through their real estate booklets for photographs that offer clear, unobstructed views of the fronts of homes. Ask each student to cut out two listings and the accompanying photographs. Have students glue their cut-outs to their construction paper as shown in the sample below.
3. Ask each student to select two features that appear in both photographs. Features might include items such as window panes, fence posts, shutters, and trees.

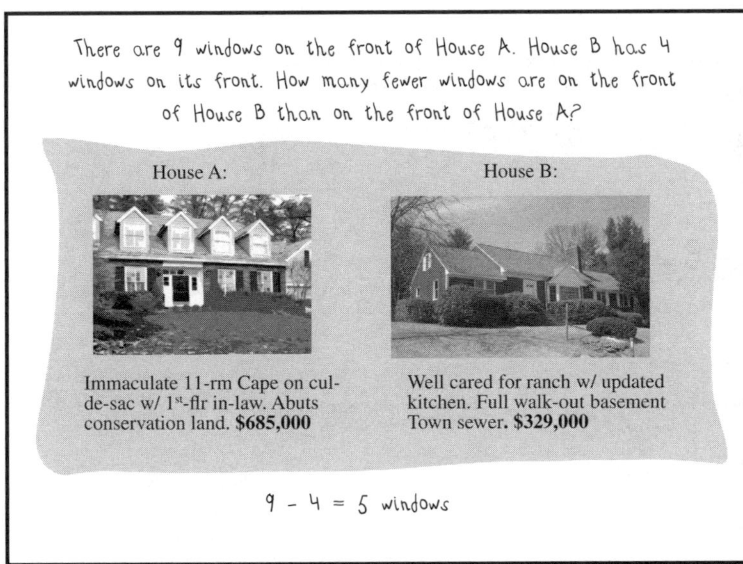

4. Next, instruct each student to write a short subtraction story problem that incorporates his chosen features above the listings. Remind students that every story problem includes a question!

5. Have each student write the number sentence necessary to solve his story problem below the listings. The difference and the appropriate label should be included.

6. When students complete all of their pages, have them order the pages from the number sentence with the lowest difference to the number sentence with the greatest difference.

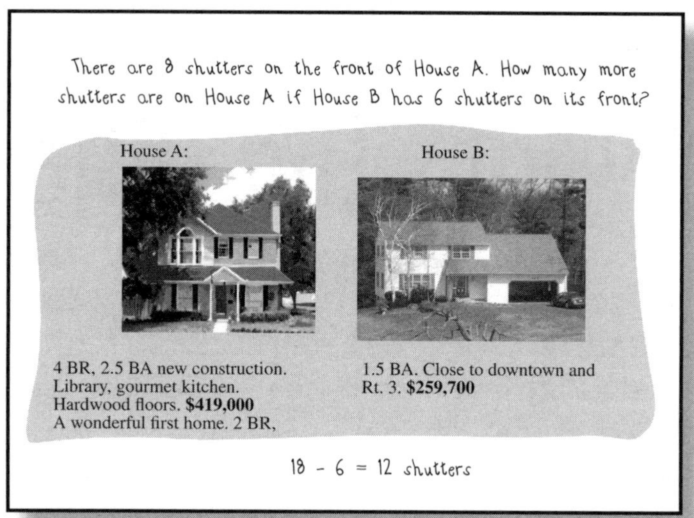

Minimum Number of Pages:

Choose a page length based on the time available and the ability level of your students. Books of five or more pages will allow you to accurately assess your students' ability to write subtraction story problems and solve subtraction number sentences.

Cover Directions:

Invite students to create neat, colorful covers that include the title and the author's full name. The covers may be decorated with the number sentence having the lowest difference and the number sentence having the greatest difference.

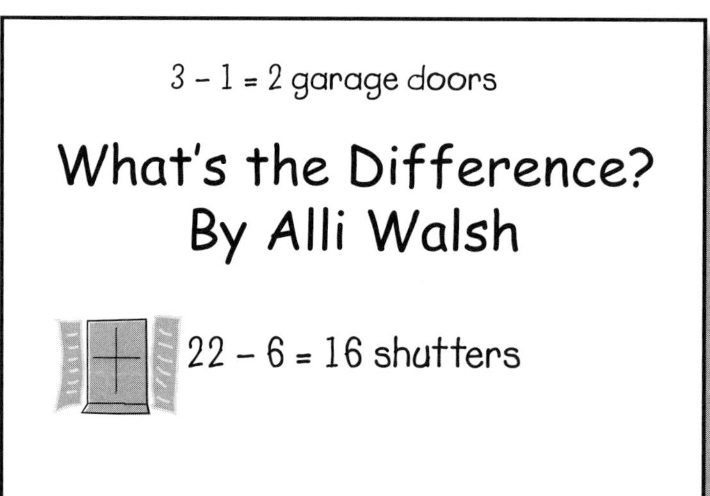

MODIFICATIONS FOR STUDENTS WITH SPECIAL NEEDS:

- Cut out and pair listings in advance to control the subtraction number sentences that the students will write. For example, you may wish to select two listings accompanied by large, clear photographs of houses with a limited number of easy-to-count windows.
- Provide each student with two listings, one marked A and one marked B. (The house with the greatest number of windows should always be used as House A.) Give students copies of the story form below. Students glue copies of their story form above their listings and fill in the correct numbers. Below the listings, have each student write the number sentence needed to solve the story problem.

There are ______ windows on the front of House A.

On the front of House B, there are ______ windows.

How many more windows are on House A than on House B?

CHALLENGES FOR HIGH-ACHIEVING STUDENTS:

- Invite each student to glue three listings to the center of the page and label them House A, House B, and House C. Challenge the students to create subtraction story problems such as:

There are a total of 19 windows on the front of House A and House B.

There are 6 windows on the front of House C.

How many more windows are on House A and House B combined than on House C?

Below the listings, ask each student to write the addition number sentence and the subtraction number sentence needed to solve the problem.

$8 + 11 = 19$ *windows* $19 - 6 = 13$ *windows*

STRAND: **Number & Operations**
SKILL: **Creating addition, subtraction, multiplication, and division number sentences**

Unlikely Listings

Students are often surprised at the large amount of information that is packed into each tiny real estate listing. So when you invite them to use their creativity and mathematics skills to turn a tiny number into something tremendous, they'll be eager to get right to work. The completed books will allow you to assess the complexity of your students' expressions and the accuracy of their computation.

Directions:

1. To kick off this activity, I begin by displaying the first line of a simple listing such as: *Move into this charming 7 room Colonial.* Together, we list the many ways that the number 7 can be expressed using one or more of the four operations. Soon, the class is ready to begin.
2. Invite the class to read the listings in their real estate booklets. Ask the students to cut out listings that contain a variety of numbers including fractions, decimals, ordinal numbers, and whole numbers. Have each student glue a listing to the top of her sheet of construction paper.
3. Challenge students to rewrite the listings, replacing the numbers—each fraction, decimal, ordinal number, and whole number—with an expression having the same value. For example, the decimal 2.5 may be changed to 5.75 – 3.25, or the whole number $350,500 may be expressed as (35 x 100,000) + (250 x 2). Listings that include measurements in feet may be converted to inches or yards.

ORIGINAL LISTING:

Last home in this **10** home subdivision! Stately **4** BR, **3.5** Bath Custom Colonial. First floor hosts **9** foot ceiling and Master Bedroom Suite. Finished game room over **26 x 26 2**-car attached garage. Professionally landscaped **.76** acre lot with circular drive. **$635,000**

Unlikely Listing:

Last home in this 6 + 4 home subdivision! Stately (12 - 10) + 2 BR, 1.25 + 2.25 Bath Custom Colonial. First floor hosts 108 inch ceiling and Master Bedroom Suite. Finished game room over (19 + 7) x (28 - 2) 5 - 3-car attached garage. Professionally landscaped 1.24 - .48 acre lot with circular drive. $127,000 x 5

Encourage students to be creative when they replace the numbers used in the listings with expressions!

4. When students complete all of their pages, have them order the pages from the highest-priced home to the lowest-priced home.

ORIGINAL LISTING:
Exceptional Colonial set on over **2** acres of exquisitely landscaped grounds. Custom built with **4900 square feet**, **4** bedrooms, **5** baths, **2** fireplaces, **3** car garage, gourmet kitchen, and finished lower level. **One** of a kind! **$1,699,000**

Unlikely Listing:
Exceptional Colonial set on over (787 - 786) + 1 acres of exquisitely landscaped grounds. Custom built with 70 x 70 square feet, 28÷7 bedrooms, 60 ÷ 12 baths, 48 ÷24 fireplaces, (4.5 - 1.75) + .25 car garage, gourmet kitchen, and finished lower level. (96-93) - 2 of a Kind! $500,000 + $500,000 + $300,000 + $300,000 + $45,000 + $45,000 + $4,500 + $4,500

Students enjoy using a calculator to check a partner's work for accuracy.

Minimum Number of Pages:

Choose a page length based on the time available and the ability level of your students. Books of five or more pages will allow you to accurately assess your students' ability to create equivalent expressions.

Cover Directions:

Invite students to create neat, colorful covers that include the title and the author's full name. The cover should be decorated with the some of the expressions that have been used in the rewritten listings.

(60 x 70) + 100 square feet
of living space

Unlikely Listings

By Finn Lyons

1.28 + (6.66 – 4.44) bath Victorian

MODIFICATIONS FOR STUDENTS WITH SPECIAL NEEDS:

- Cut out listings in advance and highlight the numbers that you wish the students to replace with expressions.
- Allow students to retype the listings on the computer. Or, instead of retyping the complete listings, allow students to simply rewrite the phrases that include numbers.
- Limit the numbers that the students are expected to replace to just the number of bedrooms and the number of baths. Prepare slips as shown below. Students glue the slips to their book pages along with their listings, adding an expression to each blank line.

________________ bedrooms	________________ baths

CHALLENGES FOR HIGH-ACHIEVING STUDENTS

- Instruct students to use at least one pair of parentheses in each expression.
- Ask students to use more than two operations in each expression.
- Tell students to use each operation a minimum of two times as the book is created. Have students convert examples of square feet to square inches.
- Invite each student to create a chart like the sample one provided below to list all of the numbers and expressions found in her book.

UNLIKELY LISTING PAGE NUMBER	NUMBER USED IN ORIGINAL LISTING	EXPRESSION USED TO REPLACE ORIGINAL LISTING
1	4 bedrooms	(144 ÷ 12) – 8 bedrooms
1	3.5 baths	(2.15 + 4.51) – (1.23 + 1.93) baths
1	5 fireplaces	(10 x 8) ÷ 16 fireplaces